ry

D0629501

Columbia, Missouri 65201

THE POPULATION PUZZLE

Overcrowding and Stress
Among Animals and Men

To the many people who helped me in the preparation of this book—I express my sincere thanks. In particular, I am indebted to John Calhoun for sending me a wealth of material related to his work at the National Institute for Mental Health, and to Mrs. Judith Lavallee for typing the manuscript.

A. H. Drummond, Jr.

The profound concern we must feel for the rapid growth of population stems precisely from the menace it brings to any morally acceptable standard of existence. . . . Without a slowing down and control of the population explosion, the life awaiting millions upon millions of this planet's future inhabitants will be stunted, miserable, and tragic.

Robert S. McNamara
President of the World Bank
September, 1970

QL
752
.D78

THE POPULATION PUZZLE

*Overcrowding and Stress
Among Animals and Men*

A. H. Drummond, Jr.

CENTRAL METHODIST COLLEGE LIBRARY
FAYETTE, MISSOURI 65248

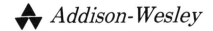 *Addison-Wesley*

AN ADDISONIAN PRESS BOOK

Text Copyright © 1973 by A. H. Drummond, Jr.
Illustrations Copyright © 1973 by Addison-Wesley
Publishing Company, Inc.
All Rights Reserved

Addison-Wesley Publishing Company, Inc.
Reading, Massachusetts 01867
Printed in the United States of America
First Printing

Library of Congress Cataloging in Publication Data
Drummond, A H
   The population puzzle.
   SUMMARY: Examines the psychological, physiological,
and sociological effects of overcrowding on man and
animals.

   "An Addisonian Press book."
   1. Crowding stress—Juvenile literature. 2. Animal
populations—Juvenile. 3. Population—Juvenile literature.
[1. Crowding stress. 2. Animal populations. 3. Population]
I. Title.
QL752.D78   301.11   72-8046
ISBN 0-201-01566-8

HA/HA

CONTENTS

INTRODUCTION: MAN IN THE POPULATION SQUEEZE

The jangle of the high-powered telephone ring cut through the background racket. The man who answered covered his free ear with a hand as he listened and then shouted replies.

"Come on, Jack," he called to his partner after hanging up, "we've got another zombie."

"Oh no," was the reply. "That's the tenth one this week. If this keeps up, we'll set a record."

The two men got up and started for the door. They threaded through the tightly crowded desks in the room, nodding occasionally to other men on duty as they passed. They weren't in any hurry. There wasn't any point in hurrying.

The office was crowded. The halls in the civic center building, however, seemed to be a solid mass of humanity. People milled about, stood in long, long lines waiting to conduct some business or other, or attempted to push their way through the crowd. It was difficult for anyone who wanted to go somewhere to make progress in the halls. The people, despite the fact that they all had business in the civic center, plugged the halls so tightly it was almost impossible to move.

Jack and his partner did their best to get to the elevator and down to the garage. They were mental health attendants. Their job was to pick up and bring into the stress ward of the hospital people who had suffered mental collapse. This happened often because the city was extremely overcrowded and people found it very hard to live. The men were on their tenth trip in five days to pick up a "zombie," their slang for a person who had gone into a complete state of mental shock. The territory they covered included just five square blocks.

Many people fought a losing battle against the noise, the foul-smelling air, the filth, the litter, and the ever-present milling crowds. When this happened, they went into a kind of walking sleep. They moved about totally unaware of the people and things around them. They did not speak. They would not eat, and could not care for themselves. This mental "disease" had become so common the city had had to take over the care of its victims.

Reaching the garage 45 minutes after taking the call, the two men got into their ambulance and headed out into the city. They had a quarter-mile trip before them. It would take the better part of two hours, however, to arrive at their destination. The city was one solid traffic jam, with pedestrians, cars, taxis, buses, and trucks all crushed together. It was like this 24 hours a day, for business, school, and cultural events continued around the clock to keep as many people as possible busy.

Four hours or so after leaving the office, the mental health attendants would sign their patient into the "quiet place." The wards where these people were cared for were the quietest rooms in the city. They were so quiet, compared to the din outside, that the doctors, nurses, and attendants felt uneasy. The lack of sound affected Jack and his partner even more. They were so accustomed to noise they felt panicky when inside the "quiet place," and couldn't wait to escape.

1
THE
OVER
POPULATION
THREAT

What you have just read is how New York, London, Chicago, Tokyo, or Los Angeles may be like after the year 2000. Does it seem strange and fearful? It should. Man today knows that one of his most important challenges is overpopulation. Yet despite this knowledge, the world is becoming more overcrowded every day.

In the United States alone, if the present rate of growth continues, the population will double in about 40 years. It will then double again in much less than 40 years. As the population soars, however, the amount of space per person goes down. People who have come to expect space around them will feel crowded. And, in the opinion of some experts, soon after they feel crowded, they will probably become uneasy, and fearful, and unhappy.

We will still be faced with air and water pollution, not enough jobs, public health problems, and

*Because there are too many people,
train passengers are packed together by "pushers" during
rush hours in Tokyo. All aboard!*

all of the other concerns that worry us today. However, these will be much more serious than they are today. Other, more pressing problems might be added to the list. There could be conflict about how to assign living space to people. Many will worry about how to get along in a crowd. We will be more concerned than ever before about how to conserve our natural resources and about how to keep from destroying the environment.

Other changes might take place. We can imagine double-decked highways, railroads, and airports. It's even possible that entire cities will be double-decked—although life will certainly be miserable on the lower deck. What will people do for fresh air, for out-of-doors recreation? The parks in today's cities are too small. What will they be like when the population has doubled? Remember that you can't double-deck a park, for plants need sunlight to grow. The United States is not alone in the population squeeze. The world's population is growing by 70 million people a year, and could double before the year 2000. What will life be like in Asia, in Europe, in Africa? Will man, accustomed as he is to plenty of space, successfully adjust to crowded conditions? Or will he fail to make this adjustment?

This book is not about overpopulation alone. It is about what man may be doing to himself by letting overpopulation take place. The problem is clear. We don't really know what extreme over-population and crowding will do to man or to his way of life. But we know enough to be worried.

As you read on, you will discover some of the things we have already learned. Life in the slums and ghettos of our large cities holds clues that we should study. The plight of an overpopulated island in the Indian Ocean will tell us much about what to expect. Perhaps even more important, studies of animals in both the wild and in the laboratory paint a dark picture. If what happens to animals when they are overcrowded will also happen to man, we are in trouble.

THE LESSON OF TODAY'S CITIES

The next time you go to a crowded party, watch how people stand in relation to other people. You will make some interesting discoveries. Notice how two people talking together stand. They almost never face each other. Rather, they hold their bodies at an angle of about 45 degrees to each other. It seems to be uncomfortable for strangers in conversation to face each other directly when they are close together.

Perhaps someone you know has the habit of pressing his face very close to yours when talking to you. Does this make you uneasy? Do you turn away or back up until the distance between the two of you is comfortable again? This happens to many people. You might notice something else. Suppose you are standing comfortably with another person, and then somebody passing behind you makes you turn toward that person. Almost immediately, he will turn away until the angle between your bodies is once again about 45 degrees. If you are sitting side-by-side on a bench, however, it doesn't seem to matter how close together you are. Aside from the physical closeness, you are at ease and can talk freely because no one is too close directly in front of you.

Observations such as these have convinced scientists that individual people and individual animals are surrounded by what is called *personal space.* That is, each individual has his own private bubble of space. This bubble bulges out around him, but more in front than anywhere else. As long as others stay outside the boundaries of this bubble, the individual is comfortable. Intrude upon his personal space, however, and he becomes distressed. If a robin's space bubble is invaded, it will threaten or flee. Man will react in terms of his culture and his needs. You or I would move away to restore our space bubble if someone came too close to us. Creep up behind a violent criminal, however, and you will probably be attacked.

Scientists believe that the idea of personal space applies to groups as well as to individuals. That is, groups of people who are somehow related or tied together set up territories. They then tend to stay within their own territories. They also resist the attempts of strangers to enter their territories.

Surprising as it may seem, such territories exist in the very worst of the slums in our cities. Here many of the people are poor, undernourished, and severely overcrowded. In some slum areas as many as six to ten people live in one run-down, cold, and dismal room.

For most of the people living in a slum, it is a trap. There seems to be no way out. The low family incomes, the high rents, the rotting buildings, the filth, and a general feeling of despair all seem to lock the people into the slum.

To the outside world, the crumbling buildings, the filth, and the sense of being trapped seem to stretch endlessly in all directions. But look closely. Something else seems to be happening. The people are living in groups. Unseen boundary lines separate these groups into territories. In Chicago's South and West sides, for example, the territories usually include two blocks next to each other.

Interesting things happen inside each territory. The sidewalks and the front steps of buildings are usually swarming with kids playing and older people talking (during summertime, anyway). They either know each other, or they know about each other by means of gossip. A woman may have friends around the corner, but none across the street because the boundary lines of her territory are in the middle of the street. Any nearby man or woman will help a hurt child, or scold one who has misbehaved. It doesn't matter what family the child comes from, for as long as he is from the territory he is safe.

Things are not the same, however, when someone from another territory moves in. Usually he will be regarded with suspicion. If so, he will also get the message quickly that he is not welcome. If someone attempts to force his way in, he may be thrown out violently.

The street gangs of the city are evidence of territoriality. Each gang has its own territory and defends it strongly against other gangs. The great adventure is to invade another territory, do something bold, and then get away. If a member of a gang is a thief, he usually will not steal within his own territory. He will probably go well outside the territory to practice his trade.

Scientists have learned that it is these territories that seem to keep things under control in the slums. People who have a sense of belonging show loyalty to their own group. While crime is a serious problem in the slums, there is no telling how bad it might be if territories did not exist.

In fact, the high-rise apartment houses now replacing the teeming slums may be the wrong answer to the problem. The old buildings are torn down, and the land is cleared. Tall new buildings are then built. Everything looks new and pretty, but trouble often sets in just as soon as the apartments are occupied.

Putting up the new buildings takes away the old neighborhood territories. This means that the people have lost their old friends and neighbors. Many have been uprooted, cast adrift among strangers. No longer do the social rules of the old territory apply. Often boys and girls run wild, out of the reach of all grown-up control. Confusion and a sense of being lost frequently set in. If people are under stress, their behavior shows it.

Whether in a new high-rise apartment or jammed into a tenement, the people in an overcrowded slum show the signs of the stresses they are living under. They are afraid and gather in small groups for safety. They often seem to be in a daze and act withdrawn. The boys and men are often in fights. They must show how tough and strong they are in order to survive.

Scientists are now very concerned about what overcrowding does to the body. The way the body works—its physiology—is affected by the conditions under which people live. For example, certain diseases and disorders are more common in the crowded slums. These conditions are partly the result of the pressures of overcrowding and a low standard of living. They include alcoholism, drug abuse, mental sickness, juvenile crime, and other problems not seen as often elsewhere.

MAURITIUS—PARADISE OR HELL?

Mauritius is an island in the Indian Ocean. In many respects it is very much like other tropical islands. It teems with people. Its principal cash crop is sugar. It has a serious overpopulation problem because serious disease—primarily malaria—was all but wiped out by the use of DDT during the late 1940's. This revolution in health, however, brought with it new problems. Today the island is so overcrowded it can no longer support itself. But its population continues to grow, along with poverty and the potential for something far worse.

Mauritius was first settled early in the 18th century. By 1833 the population had increased to 100,000. Originally the island was regarded as a healthy place to live, but its growing population brought problems of sanitation. Rats carried in plague and other diseases as they came ashore from ships. Epidemics of smallpox, cholera, influenza, and bubonic plague swept the island. Malaria became a serious problem later in the century. Eventually it emerged as the island's chief cause of death. During these years death from disease kept the population under control.

From 1900 to 1915 there were more deaths than births. The disease problem continued unchecked until World War II, when DDT came into use. At this point births began to exceed deaths. Deaths from malaria dropped from 3,534 in 1945 to just three in 1955. During the mid-forties 155 babies in every thousand died. This number dropped to 62 deaths per thousand by 1961. These were dramatic gains in the health of the island's population. As notable as they were, however, they speeded up the overpopulation problem.

As early as 1962, it was known that the island was approaching the limit of its ability to support the population. At that time almost half the population was either under 15 or over 65 years of age. These people required care. They did not contribute to the island's economy—that is, to its ability to produce goods and services that could be sold. The working half of the population, on the other hand, was finding it more and more difficult to produce the food and goods needed to support all of the people on the island. In short, the population had outstripped the island's ability to support it. But the population continued to grow. It still does today.

When an island economy cannot support its people, it must import what it needs to survive. Moreover, it must pay for these imports. In Mauritius' case, the one cash crop available has been sugar. But not enough sugar can be grown to pay for the needed imports. Thus the people of Mauritius, isolated as they are in the Indian Ocean, cannot support themselves. Despite this unhappy situation, however, their numbers continue to grow. The result is an increasing population trying to live on a fixed economy. Put another way, when opportunities for business and economic growth stop and population growth continues, the amount of money available for each person decreases. Experts say that the per capita income has declined. What this means very simply is that each family has less and less money as time passes. There is less money to pay for such important things as medical care, housing, transportation, food, and so on. For the island of Mauritius, there seems to be no escape from this trap.

The government has attempted to improve the island's economy. It has also taken steps to improve the educational system. These efforts, however, have not been very successful. Today life on

Mauritius is all but unbearable. Many thousands of people are living on a near starvation diet. They suffer from severe malnutrition, with little hope that conditions will improve. In addition, they are very crowded. The island has one of the world's highest population densities—about 900 people per square mile.

What is worse, at its present rate of growth, the island's population will probably double in about 20 years.

The government has tried also in recent years to bring population growth under control. It knows that with a limited economy, the island can support only a limited number of people. Family planning, that is, controlling the number of children born, seems to be the answer. Unfortunately, family planning has been slow in catching on. More and more children grow up with no hope of ever working at a job, with no hope of ever supporting themselves or the children they in turn will bring into the world.

Some experts have predicted what will happen if this problem is not solved, and soon. They foresee the need for martial law. There will be so many people only military force will be able to keep them under control. In addition, the experts predict that the government will have to set up special camps for the people who cannot find work.

One expert goes further. He thinks the working public will demand that the camps be fenced in; that the unemployed poor be locked up permanently. Imagine what this would be like. People would spend their entire lives behind barbed wire. They would be born, grow up, and then die in a prison camp, never to work and never to know what the outside world is like.

This is a horrible example of what might happen if population growth is not controlled. Even more disturbing, it could happen, and soon, on the island of Mauritius. This small piece of land in the Indian Ocean is like a miniature world. What happens to it could happen to the entire planet as well. Both are islands, one in an ocean and the other in space.

Is it possible for man to become so numerous the planet can no longer support him? Most experts answer "yes" to this question.

Even if the planet could support them, what will happen to people when they are unbearably overcrowded? Man is a creature who evolved to his present form with lots of space to move around in. How will he react if this space disappears? Will he adapt quickly and learn how to get along in a crowd? Will an individual's space bubble become smaller as less and less room is available? Or is there a limit? Will man learn to feel joy, happiness, and fulfillment in an overcrowded world? Or will he end up like the "zombies" in the imaginary crowded city described at the beginning of the book?

The answers to these questions are not known today. Clues to the answers, however, have been discovered. They range from knowledge of the body's defenses against stress to the results of studies of animals in the laboratory and in the wild. When all of the pieces in the puzzle have been put together, we will know more about man's chances on an overcrowded planet.

2
SPACE
AND THE
CITIZEN

Nearly everyone has watched a lion tamer at work in a circus. With a chair or stool in one hand, a whip in the other, and a gun at his belt, he enters the cage with the lions. Then, moving toward one of the lions slowly and carefully, the lion tamer cracks the whip and raises the chair. He seems to signal what he wants the lion to do by waving the chair. The lion backs up obediently until it reaches the bars of the cage. Then it turns, growls its anger and defiance at the man, and leaps up on a stool.

Marvelous! The lion tamer, using his greater intelligence and sheer will power, has "tamed" the king of the jungle. He has not only kept himself from being attacked, he has forced the lion up onto a stool—a silly perch for the king of the jungle.

ANIMAL SPACE REQUIREMENTS

But this isn't what happened at all. The lion tamer knows something the audience does not know. He knows that when he enters the cage the lion will flee—it will attempt to get away. This is the first part of the act, when the lion is moving away from the man toward the far side of the cage. What is happening is quite simple. The lion tamer approaches the lion until he reaches the animal's *flight distance.* When a man or a potential enemy reaches the flight distance of a wild animal, the animal turns and flees.

Each animal has its own flight distance. Baboons in the wild, for example, sometimes flee when a man comes within 100 yards. For an antelope, the distance can be as great as 500 yards. To truly tame a wild animal, flight distance must be reduced. The animal becomes a pet, of course, when flight distance has been reduced to zero.

But now back to the lion in the cage. It is trying to escape because the lion tamer has reached its flight distance. As the man slowly approaches the lion moves away until it is backed up against the bars of the cage. At this point another factor enters. This is the lion's *critical distance* or *zone.* The critical distance lies between flight distance and *attack distance.*

The lion is backed up against the bars. It can no longer run away. If the lion tamer now comes closer, he enters the lion's critical zone. This causes the cornered animal to turn and begin

How is the tamer controlling these lions?
With a whip? With his shouts?
Or by his knowledge of the lions' space requirements?

PHOTO COURTESY OF RINGLING BROS.—BARNUM AND BAILEY COMBINED SHOWS, INC.

stalking the man. The lion snarls, roars, and whips its tail in anger. It is preparing to attack the man. It will stalk, however, until the man reaches its attack distance.

This is where the lion tamer outwits the lion. He permits the animal to move along the bars until a stool comes between them. At this point the lion, getting ready to attack, will climb atop anything in between itself and the lion tamer. It leaps up on the stool, ready to attack. But the man steps back at just this moment, out of the lion's critical zone. Instantly the lion stops stalking. It sits down on the stool. The man, its enemy, has moved outside its flight distance also. There is no longer any need to either flee or attack.

This story is one example of how the space requirements of an animal can sometimes work. Each different wild animal has its own flight distance and critical zone. As you can see, these distances mark the boundaries of space bubbles around the animal. The effect of these bubbles, or space balloons, is to maintain spacing between animals. When space bubbles overlap or are invaded, the animals they surround become uneasy. They may take flight, or perhaps attack the intruder.

But there is more to the use of space by animals. Some huddle together; they prefer to touch, to stay close to each other. Others avoid touching; these animals always maintain a certain distance between themselves. Pigs and parakeets are contact creatures. They love to huddle together. Dogs, cats, horses, and man are noncontact creatures. They prefer to keep a certain distance apart from their fellows. Thus, many animals as well as man use space two different ways. They set up territories that they defend, and they keep a certain distance between individuals.

MAN AND SPACE

Man's long history of living and working together has just about eliminated flight distance and critical distance from his reactions. Other recognizable distances, however, do seem to be used by people as they go about their lives. These can be broken down into four basic groupings—*intimate distance, personal distance, social distance,* and *public distance.* Intimate and personal distance refer to close contact between two people or small groups of people. Social distance and public distance come into play when closeness is not desired —attending a class or shopping, for example.

People react differently when they come into contact with others. Sometimes even the same person will behave differently in two separate, but similar situations. This is the result of how we have learned to expand and contract our space bubbles. For example, the violent criminal mentioned earlier cannot bear to have a strange person creep up behind him. His reaction is to turn and attack when his space bubble to the rear has been entered. But suppose he knows and trusts a person completely. This person can usually draw near to him from behind without fear of attack. For someone he trusts, the criminal's space bubble to the rear is smaller.

Let's look a bit more closely at the distances used by man. Remember, however, that we are talking about today's man. It is interesting to try to figure out how these distances might change if the world should become as crowded as the imaginary city at the beginning of the book.

Intimate distance is the distance of comforting and protecting, as when you hug and soothe a child who has been crying. It is the distance of a football block or tackle, and of a wrestling match. It is also the distance of showing love between a man and a woman. At intimate distance the heat of another person's body is noticeable. So is the odor of his body and his breath. At intimate distance the fine details of a person's skin, hair, and eyes are clearly visible.

Most adult Americans cannot stand the closeness of intimate distance, unless there is a good reason for it. This is especially true in public places, where the use of intimate distance is definitely frowned on. But, people do find themselves at intimate distance against their will.

This often happens on a crowded bus or elevator. Watch what people do in these situations to take the intimacy out of intimate closeness. They stand perfectly still to avoid touching their neighbors. If a leg, the body, or an arm or hand should touch someone else, it is pulled back immediately. If it is impossible to avoid touching, the muscles of the body are kept tense. How one sees is affected, too. The eyes stare into the distance, for it is thought to be improper to look directly at a stranger at intimate distance. Only glances are allowed.

Intimate distance in man extends from touching to about 18 inches. Personal distance then takes over, and ranges from 18 inches to about four feet. The term personal distance was first used to describe the space noncontact creatures keep between themselves. Watch birds perched on fence rails, rooftops, or wires for a good example of personal distance. A separate distance exists for each different type of bird. For gulls, for example, the distance is about one foot. Swallows, on the other hand, seem satisfied with six inches.

Among people, personal distance seems related to what they can do to each other by reaching out.

Showing friendship by placing your hand on another person's shoulder, but at arm's length, is an act that takes place at personal distance. How far you stand from another person often tells what you mean to that person. A boyfriend or girlfriend will be quite comfortable standing within your personal zone. If a stranger should stand that close, however, you will probably be ill at ease.

At the outer limits of personal space two people can touch fingers if they extend both arms. This is about the limit of physical control. Beyond this point another person is "out of reach." People sometimes do strange things at this distance. For example, if they can't avoid being close to other people, but want to feel that they are still at arm's length, they may wear dark glasses all the time. In a crowd, especially indoors, this says, "You can't reach me. You are not close to me. I watch you, but you can't watch me."

Social distance in man extends from four to about 12 feet. This is the distance used when dealing with people other than very close friends, sweethearts, or one's husband or wife. It is the space we keep between us when making new friends at a party, or when doing business in an office.

The eyes are very important at social distance. Two people at this distance are far enough apart to break off by losing eye contact. Thus, when talking at social distance, people attempt to hold each other's eyes. It is almost impossible to say something directly to another person without catching his eye first. This is why people will often crane their necks, lean one way or the other, or even shift position when conversing at social distance. They are moving to avoid objects that come between them. If they didn't move, eye contact would be broken, and communication would stop.

Public distance is outside the circle of personal involvement. It extends from 12 feet to 25 feet or more. From about 12 feet a person can flee from or evade a threatening action from another person. Thus, public distance in man may include the almost lost flight distance that is still seen in wild animals.

At public distance people are more formal in their communications with others. The voice is louder than when at closer distances. More carefully structured sentences are used. This is the distance between a teacher and his class, although the first row of students may be closer to him than 12 feet. The loudness of his voice, its tone, and the way he speaks all indicate that he is at public distance. Note the change that takes place, however, when the class period is over and a small group of students gather around the teacher. Now his voice is softer, his style is more informal, and he looks directly at the student he speaks to. The group is now at social distance. Each person's manner changes when he moves from public distance to social distance.

It has only been recently that man's space needs have been thought of in terms of the sensitive space bubble that surrounds every individual. Prior to this it was thought that a person occupied only the space filled by his body. Now we know better. A person's body is at the center of his space bubble. He feels and experiences at its outer edge, however, not where his skin ends.

This is an important fact, one that city planners, architects, and even school teachers should remember. People can be cramped by the spaces in which they must live, work, and learn. Even worse, if they are overcrowded and feel stress, they may say and do things they would not do if they weren't crowded. Thus, the high rate of crime in the slums, the mistakes made by a clerk in a crowded office, and poor discipline in a classroom may be the result of crowding and stress. These acts may be something a person cannot control.

What will happen to people if the earth's population growth is allowed to continue unchecked? If conditions become like those described at the beginning of the book? Will we adapt and learn to live happily in the midst of a tremendous crowd? Or will our space needs prevent this from happening? It is ironic that man's tremendous population growth has caused the extinction of many other animals.

He could do the same thing to himself.

3
GUPPIES, VOLES, AND RATS IN THE LABORATORY

Odd as it may seem, the science of man's space needs has its roots in economics. The story goes back to 1798, when the English economist Thomas Malthus published his *Essay on The Principle of Population.* Malthus' idea was basically simple. He proposed that human populations increase at a faster rate than food supply increases. Sooner or later the growing population catches up with food supply. At this point there isn't enough food to support any more people. As a result, the population stops growing. The maximum amount of food available is what controls the population—the number of people in a country, on a continent, or on an island.

Two of the world's greatest biologists, Darwin and Wallace, accepted Malthus' idea as a law of nature. From it they concluded that all living creatures compete for food—a limited resource. But because the food available is limited, only the strongest or most fit survive to produce young. This in part is the basis of the original theory of evolution. Living creatures evolve because the most fit are selected through competition. Only the strongest—the ones best able to get along in the environment—grow up to mate and produce young. The result is a gradually changing and improving population.

Today we know that Malthus was wrong. A limited food supply does not always mean that a rapidly increasing population will stop growing. As we shall see later, there are other reasons. Thus, Darwin and Wallace made an error in accepting the Malthusian idea. Nevertheless, their theory of evolution stands as one of biology's most important ideas. They were correct in stating that living creatures evolve. They did not, however, correctly describe the details of how evolution takes place. Today we honor these men for their tremendous insight. In a very real way, they provided us with the basis for the golden age of biology.

If the amount of food available is not what controls population growth, what is? Scientists have observed over and over again that many animals in the wild somehow keep their numbers in check. In almost all cases, animal populations stop growing long before they reach the upper limit of food supply. Why does this happen? As we shall see, there is no clear-cut answer to this question. But now let's go back and look at some of the experiments that have provided clues to the solution of this puzzle.

SOME PUZZLING OBSERVATIONS

During the early 1930's two biologists, C. M. Breder and C. W. Coates of the New York Aquarium, performed an unusual experiment with guppies. These little fish multiply very rapidly. A pregnant female guppy, for example, may produce as many as three broods, born about 28 days apart. In each brood, two females are born for each male produced.

In the experiment, two tanks of equal size were set up. Each had ample food and was aerated to the point that many, many fish could live in the tank. In one tank the scientists placed one heavily pregnant female. In the other they placed 50 guppies, but in a combination not found in wild populations. This tank contained one-third adult females, one-third adult males, and one-third babies.

The results of this experiment astonished the two scientists. The single pregnant female produced large broods right on schedule. At the end of six weeks, however, there were just nine fish in the tank—six females and three males. All of the surplus young had been eaten. In the other tank a rapid die-off started just as soon as the tank was stocked. All new babies born were eaten immediately. As in the first tank, at the end of six weeks only nine fish remained—three males and six females. What had happened? There was more than enough food, and the water was aerated far beyond the needs of the fish. Clearly, some other force was at work in the tanks.

Many years later an experiment by C. J. Krebs and K. T. DeLong with California voles produced similar puzzling results. Voles are small rodents closely related to lemmings and muskrats. In this experiment, two areas of land of about one acre each were set aside for study. One of the areas contained a small population of voles—five males and eight females. The vole population of the second area, however, was increasing naturally at a rate of about three percent a month.

The experimenters were interested in the area containing the five males and eight females. They fertilized the area to obtain a rich plant growth. They scattered rich food for the voles. As expected, the population boomed. It reached a springtime peak of 47 males and 53 females. But then—surprisingly—by August the number of voles had dropped back to exactly the original number of animals—five males and eight females. Once again, some factor other than food supply prevented population growth.

These and other experiments have shown scientists that animals in the wild possess some means of regulating their population growth. They don't simply increase in number until there is no longer enough food. Why does this happen? The answer seems to lie in the space needs of the animals. John Calhoun's famous experiment with rats will help us understand this problem.

CALHOUN'S RATS

In Calhoun's first experiment with rats during the early 1950's, he placed five pregnant wild Norway rats in a single quarter-acre outdoor pen. Fifty thousand descendants could have been produced by these five females during the 28 months of the experiment. Moreover, if these rats had been kept in eight-inch cages, the 50,000 animals could have gotten along nicely. If the cages had been two feet square, 5,000 rats could have lived in health on the quarter-acre.

There were no cages, however. There were also far fewer rats. At no time did the population of the quarter-acre pen go higher than 200. It eventually leveled off at about 150, and stayed at that figure.

Calhoun wondered why the permanent population stayed at about 150.

Studying the behavior of the rats, he noticed some unusual things. For example, even with 150 rats in the pen, there was so much fighting the females were not very successful in raising their young. Very few survived. Calhoun noticed also that the rats had organized themselves into little colonies of about 12 rats each. Twelve seems to be the largest number of rats that can get along in a group in the wild state. Calhoun observed something else. Signs of stress appeared in these small bands of animals. Behavior changed; there were even physiological changes in the animals' bodies.

Calhoun concluded that space requirements kept the population down at the 150 mark. Even though there was plenty of food and shelter, the rats' personal space requirements and their need to stake out and defend territories prevented any further population growth. These conclusions led Calhoun to the now famous experiment carried out inside his barn.

These experiments with guppies, voles, and rats, as well as others, seem to say that animals in the wild regulate their population numbers in order to avoid overcrowding and stress. Recognizing this, Calhoun set out to discover what would happen if rats were forced into an overcrowded condition.

In an early Calhoun experiment, male rats harassed a female (center) until she retreated into her burrow.

He built a series of pens that could be observed without disturbing the rats. Each group of pens contained four separate but connected pens. These were connected in the same manner as four railroad cars. That is, the two end pens opened onto the central pens but not to the outside. Thus, the two middle pens had two entrances each. The end pens, however, had but one entrance each.

Each of the pens was equipped with everything a rat needed to live a full and healthy life. Moreover, each pen was large enough to comfortably house 12 rats, the number Calhoun had noted in the bands of wild rats.

At the beginning of the experiment, one or two pregnant females were placed in each of the pens. The young were allowed to grow up and roam freely in all four pens. As the population grew, however, the rats began to be crowded. Soon strange forms of behavior began to appear.

At first a status struggle took place between the males. As a result, the two most powerful males took over the two end pens—one each. Each guarded his territory by standing watch at the entrance. At first, when the rats were roaming freely in all four pens, there were several males and females in each of the end pens.

The two dominant males quickly set up harems in their territories. They allowed females to come and go at will. When other males left, however, they were not permitted to return unless they accepted the dominance of the territorial leader.

Life for the rats in the end pens was reasonably good. The females belonged to the dominant male. They were good mothers and built decent nests. They took reasonably good care of their young and raised about half to the weaning stage. Any other males that visited these territories—several slept there, for example—stayed completely away from the females. They did not even attempt to mate with females in season. Thus, the end pens showed the social organization typical of wild rats.

Conditions in the middle pens, however, were not so pleasant. As overcrowding increased, several different types of antisocial behavior appeared. There were many dominant males, for example. These males could not, however, organize territories because there was no way for them to guard two entrances at the same time. Fights broke out frequently. Thus, victory or defeat determined the male rat that would be "king-of-the-hill" or at the bottom of the heap. These positions changed often.

As time went on and overcrowding became more severe, Calhoun was able to recognize several new male roles. There were completely passive males that took no interest in either fighting or mating. These were the drop-outs (the "zombies"?) of the middle pens. Another group Calhoun called "the probers." These active males spent their time in gangs chasing and attacking females. They abandoned normal rat courtship and mating patterns, and performed what can only be called "rat rape." These males showed no interest in the status struggle, yet they were the most active males in the pens. In essence, they were the criminals of the rat population.

Another group of males was pansexual. For them, any sex partner would do. Any rat in season (or not) was regarded as a potential mate. Finally, a group of males withdrew completely from all social and mating activity. These animals roamed throughout the pens when the other rats slept. These were the "loners"—the "outcasts." They had nothing to do with other rats, but went on the prowl when the rest of the colony slept.

Females and the young suffered far more than males in the middle pens. Female Norway rats are usually good mothers and build decent nests. The females in the harems of the end pens continued to do this. Those who lived in the middle pens did not. Their nests lost quality rapidly. Toward the end of the experiment no nests at all were built in the middle pens.

Care of the young suffered as a result. With poorly built nests and eventually none at all, the

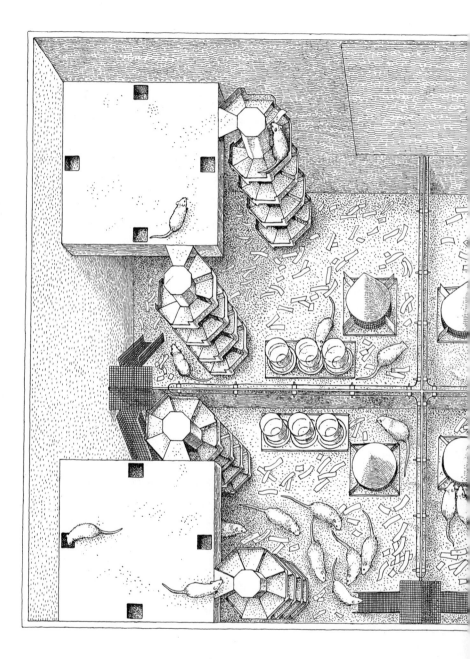

In Calhoun's experiment,
the behavioral sink
developed in the
two front pens.
Dominant males—shown
next to the ramps
in the two rear
pens—guard their
territories against
intruders.

young became scattered at birth. Many were abandoned, some were eaten, others were trampled on. Few survived. Infant deaths in the middle pens reached 96 percent. The females themselves had a high death rate. About half died following sexual attack or sickness during pregnancy.

Calhoun invented the term *behavioral sink* to label what went on in the middle pens. A behavioral sink is a place where undesirable behavior takes place. There is no question that what took place in the middle pens was extremely unusual. Almost all of the normal behavior of Norway rats disappeared, to be replaced by what was often extremely antisocial behavior.

Calhoun believes that the sink is "the outcome of any behavioral process that collects animals together in unusually great numbers." He goes further. He found that the rats enjoyed the behavioral sink. Those from the end pens were strongly attracted to the social excitement of the middle pens. They even preferred to eat there, rather than in the protected end pens. Only three percent of the end-pen harem females could resist the excitement of the crowd.

What does this experiment suggest to us about overcrowding in man? Will crowded human populations behave the same way as the rats in the middle pens? Will our major cities, increasing in population every year, become behavioral sinks? Or has it already happened in the crowded slums? We know now that in rats, overcrowding upsets and destroys important rules of behavior. This is followed by disorganization, and eventually the deaths of many animals. This is called population collapse or die-off.

Will it happen to overcrowded human populations? After reading about Calhoun's experiment, you may want to agree. Be careful, however. It is dangerous to leap to conclusions when animal and human behavior are compared. Let's look at some additional experimental work. We may just have a surprise or two in store for us.

4
CROWDING
OR
CONFLICT?

Calhoun's experiment suggests that overcrowding produces a behavioral sink. This may not, however, apply at all times to all animals. About ten years ago Calhoun and another biological scientist—Alexander Kessler—worked together briefly to help Kessler set up an experiment with mice. Kessler wanted to study how population growth affected the survival of mice with differing genetic backgrounds. An animal's genes, the tiny molecular units that carry hereditary characteristics from parents to young, are what make up the genetic background.

Kessler asked Calhoun to help him design pens for his mice. This Calhoun did. Unfortunately, however, the two men stopped corresponding before Calhoun had given all of his suggestions. Undaunted, Kessler went ahead with his experiment. It produced a very unusual result. This result was so unusual Calhoun said later he was pleased that correspondence between the two men had broken off. If Kessler had taken all of the advice Calhoun had to offer, the experiment (and its result) would have been much different.

EXTREME POPULATION DENSITY

Kessler did two things that Calhoun would not have done. First, he built his pens—two of them—about one-quarter the size that Calhoun would have recommended. In a way this was a happy accident, for Kessler had only a limited amount of space for his pens. Second, not knowing what Calhoun had found out about animals living in small groups, Kessler stocked each pen with 32 breeding adults. This was far more than Calhoun would have started with, although it was the correct number for the genetics studies Kessler wanted to do.

The result was totally unexpected. At the end of the experiment one pen contained over 800 mice. The other was supporting over 1000 adult and weaned animals. In the case of the second pen, conditions were so crowded there was less than three square inches of space for each mouse. Never before had an experimental study using rodents led to such population density. In all other similar experiments population numbers had leveled off at about one-tenth the total observed here. What had happened? Why were these mice able to get along reasonably well at a population density ten times as great as any ever before observed?

There are no answers to these questions; that is, answers with solid experimental evidence backing them up. Calhoun has given a great deal of thought to this result. He thinks he knows what caused it, and has set up a large scale experiment to test his explanation. More about this experiment later.

He points out that there was really only one difference between Kessler's experiment and those of other scientists. Kessler started his experiment with a much larger number of colonizing animals. This suggests to Calhoun that the more animals there are to start, the larger the population will be before stress produces a behavioral sink. Put another way, if the animals in a colony are very crowded right from the beginning, they never have a chance to establish territories or become accustomed to lots of space. As a result, they do not feel the pressure of crowding as early as the animals in the other experiments.

Suppose we start an imaginary experiment with mice that uses two pens. Both pens are the same size, and contain the same provisions for food, water, and nest building. In one we place two mice, a mature male and a mature female. The other we stock with 16 mice, eight males and eight females. As time passes, the two populations grow. Now suppose that at some point in time each pen has 100 mice. If Calhoun's idea is correct, there should be marked differences between the behavior of the two populations.

The mice in pen one, the population that started with two animals, should show the effects of stress much more than the mice in pen two. The pen-one mice are descended from animals that had plenty of room to roam about in. These mice established and defended territories. They became accustomed to space, and felt pressure when population growth cut down on their space.

In the other pen the mice were crowded right at the start of the experiment. There was no way for dominant mice to set up and defend territories. Also, because of the crowding the personal space available for each mouse was very small. As a result, far less stress developed as the population in this pen grew. There was very little aggressive behavior. And finally, personal space was so reduced it was limited to roughly the boundaries of the individual mouse's skin.

Suppose this idea of Calhoun's is correct, and that like the mice in Kessler's experiment man can learn to get along under severely crowded conditions. The people of Mauritius and perhaps the crowded populations of cities such as Calcutta, Hong Kong, and Tokyo seem to suggest that he is right. If experiments now under way give this theory strong support, the world's overpopulation problem would become even more frightening.

Experts have always felt that man's space needs will prevent overpopulation to the point of "standing room only." Kessler's experiment, however, suggests that this is not so, that nothing stands in the way of standing room only conditions on the planet. Can you imagine what it would be like if there were only one square yard of space per person throughout the earth?

Man does have space needs. It may also be true, however, that man is so adaptable he can learn to get along with less and less space. If this is so, no natural force will curb human population growth. Only deliberate action by society will do the job.

STRESS FROM CONFLICT

We have seen that the experiments of Calhoun and Kessler do not agree on the effects of crowding. It is therefore difficult to apply these results to man. In addition, there is another factor that must be taken into account. This is the matter of harmony among the people in a group. If there is disagreement and conflict within a group, will

DRAWING BY WEBER; © 1971; THE NEW YORKER MAGAZINE, INC.

"Excuse me, sir. I am prepared to make you a rather attractive offer for your square."

there be a feeling of overcrowding with fewer people? If there is harmony, on the other hand, can more people work and live together before the group feels overcrowded? Another Calhoun experiment throws light on this problem.

In this experiment, 16 rats were living together in one small pen. The rats in the group got along with each other very nicely. There was no evidence of stress, no indication of overcrowding. These rats, however, had to learn a task in order to survive. They had to cooperate in pairs in order to obtain water to drink. Here's how this system worked.

A specially constructed water container was placed in the pen. The front of this container was rigged with two levers. The levers were then separated by a sheet of plastic that the rats could see through. In addition, a wired mat was placed in front of each lever. This system was set up to deliver a drop of water to a rat only when there were two animals in place in front of the levers. With a rat standing on each mat, pressing either lever would deliver water to drink.

The 16 rats in the pen learned this task very quickly. Whenever a rat felt thirst, it would stand in place in front of one of the levers. Then, as soon as this rat was noticed, another rat would take the other place. The two rats would then approach the

Two cooperating rats approach a water lever—similar to the one used in Calhoun's experiment.

levers together. At this point, with a rat standing on each mat, the thirsty rat could get a drink by pressing its lever.

These rats had learned to cooperate with each other. They had established a system of values. In a very real way, they had learned that it was necessary to be considerate of their fellows. Without this concern, it would have been impossible to obtain drinking water.

Calhoun set up a similar colony next to this pen. The two pens were separated by a wire fence. This second pen also contained 16 rats. There was one important difference, however. In this second pen, it was possible to obtain water only by approaching the levers alone. If there was a rat on each of the wired mats, no water could be obtained.

These rats also learned quickly how to get water. They avoided each other at the drinking station. This, however, produced a new problem. The rats spent about three quarters of their time sleeping. As a result, there were often too many thirsty rats for the drinking station to take care of.

Rats, as you know, are very clever animals. They are capable of coming up with unusual solutions to many problems. In this case, one of the rats having trouble getting water in the second pen learned to jump the fence between the pens. Once over the fence, this animal headed straight for the drinking station. It did not know, however, that the rules for getting water were different in this pen. Serious trouble started immediately.

Each time the visiting rat approached a water lever, one of the "home" rats would join him on the opposite side. This, of course, unlocked the levers so that water could be released. The visiting rat, however, had learned that to obtain water it had to go to the lever alone. It therefore resented the presence of the home rat. Its reaction was quick and to the point. Each time another rat entered the water channel with it, the visiting rat backed out, grasped the other animal by the tail, hind feet, or skin, and dragged it out. Many of the home rats, trying to help an apparent companion get water, were quickly bruised, cut, and seriously upset by this strange behavior.

The effect on the colony was striking. This new stress was so upsetting eight of the home rats died shortly after the visits of the outsider began. The remaining eight rats became so upset and unhappy Calhoun had to put them out of their misery.

Do these results apply to people? Again, we must be cautious and look for more evidence before drawing any conclusions. Suppose, though, that something similar does happen to people. It might help explain the increased conflict seen around the world in recent years.

First, we know the planet's population has been growing rapidly. Second, high speed air, sea, and land transportation has made it possible to get to almost any part of the world very quickly. And third, there has been a tremendous increase in the amount of information available about other cultures.

The net effect of these changes is that people are being exposed more and more to the values of other lands. But these values often conflict. We have seen the result of this sort of conflict between the inner-city ghetto and the suburbs, between

Arabs and Israelis in the Middle East, and between Americans and Vietnamese in Southeast Asia. It is not a pretty picture.

Will this type of conflict become more common as population growth continues? As people from different cultures come into contact more often? People are far more complex than rats. Moreover, Calhoun's experiment was quite simple in its design and purpose. Experiments are underway to get better answers. In the meantime, we should be aware of this important aspect of overcrowding. When cultures with different values come into contact, violent conflict may result.

5
ANIMALS
IN THE
WILD

Many people are critical of laboratory experiments such as those just described. They say that putting animals in cages creates conditions so unlike the wild state that experimental results cannot be trusted. Calhoun, Kessler, and the others working in laboratories would probably say "Yes, but . . ." to this criticism. Their purpose, after all, is to study how animals react to unusual environments, and then to compare these results to the behavior of wild populations. They are interested, too, in how these results may apply to human populations.

In any event, there are many good studies of animals in the wild state that have something to say about the problem of crowding and stress. Thus, to satisfy the critics, and also to broaden our understanding of the problem, we will look at several of these studies.

KAIBAB MULE AND NEW ZEALAND RED

Late in 1906, President Theodore Roosevelt created the Grand Canyon National Game Preserve. This million-acre area roughly outlined the Kaibab North Plateau in northwestern Arizona. It was home to some 3000 Rocky Mountain mule deer, an animal that had lived there for centuries.

Roosevelt instructed the U.S. Forest Service to give high priority to the "propagation and breeding" of mule deer on the new game preserve. The foresters went at this task enthusiastically. Scientific wildlife management, however, was in its infancy at the time, and just about every decision the foresters made turned out to be a serious mistake. Very little was known about managing deer. Even less was known about ecological relationships— that is, about how the living creatures of a region interact with each other and with the environment to produce a so-called "balance of nature."

The foresters' first step to protect the deer was to ban all hunting. In imposing the ban, however, it was completely forgotten that Indians had for centuries taken about ten percent of the deer herd each year without damaging it.

The second step was a declaration of war on the predators in the game preserve. The region teemed with coyotes, mountain lions, and bobcats. There were also a few gray wolves. Like the Indians, these predatory animals had been on the Kaibab Plateau for centuries. We know now that a balance must have existed between the preyed-on deer and the predators.

This balance, however, was disrupted during the next 25 years. Some 5000 coyotes, 800 mountain lions, 550 bobcats, and 20 wolves were killed by the foresters during this time period. Many others were killed by private hunters.

At first, this experiment seemed a great success. The number of deer increased with each breeding season. In just 12 years the deer population had grown to four times its original number. But signs of danger were also beginning to appear. The range was being overbrowsed; it began to deteriorate rapidly. Forage grasses died out and woody plants were killed as the deer cropped them too severely. Widespread erosion developed because the roots of dead plants could no longer hold the topsoil in place.

By 1923, according to unofficial count, there were about 100,000 deer on the range. Alarmed at this point, the Government took steps to reduce the deer herd. Unfortunately, these efforts were both unsuccessful and too late. The damage had been done. The deer population had far outgrown the carrying capacity of the range—destroying its own supply of food.

The result? Deer by the thousands froze to death or starved every winter. There was just no way the range could support the herd. To make matters worse, each spring the winter losses were largely replaced by new fawns. This cycle then repeated itself year after year.

The Kaibab North Plateau, once a lush natural wilderness home for Indians, mule deer and their predators, and countless other animals, had become a ravaged and barren desert. Only the protected mule deer remained. But these animals had become a wretched shadow of the original deer. They were thin and gaunt from starvation, and riddled with parasites. Other forms of disease further damaged their health.

This disaster occurred in 1923. Strong corrective measures eventually brought the herd's population under control. Later, however, when a better balance existed, further game management mistakes were made. The result was another large scale population crash. This second die-off took place in 1954.

Man's mistakes with the Kaibab mule deer have fortunately now been corrected. Also fortunate, we learned two very important lessons about game management. The first is that grazing or browsing animals must not be protected from their natural enemies. To keep a plant-eating herd in balance with its food supply, predators must be protected as well. The second lesson is that man must take control of the environment himself if he upsets the natural balance between plants and plant eaters and between predators and prey. With all natural controls on population growth gone, only man has the resources to step in and prevent disasters such as the one described here.

Experiences in other parts of the world bear out these lessons. For example, many years ago in New Zealand the large red deer was introduced in the complete absence of predators. As a result, the deer population rose sharply to a peak. This was then followed by a massive die-off. After this first die-off, the population rose and fell periodically on a somewhat smaller scale. When this population was studied in 1958, some 60 years after the deer were introduced, the environment was found to be severely damaged. Moreover, the deer on the range were in very poor condition, showing that a further die-off was to be expected.

There is no suggestion anywhere that the stress of overcrowding had anything to do with the Kaibab and New Zealand die-offs. Both of these cases, however, go back to the turn of the century. This was long before scientists became interested in the effects of overcrowding. Nevertheless, in both cases the population soared far beyond the ability of the environment to support it. It seems reasonable to feel that the stress of overcrowding, as well as starvation, had something to do with both the die-offs and the very poor condition of the animals. As we shall see later, there is strong evidence that this is the case.

SELF-CONTROL OF POPULATION: AFRICAN ELEPHANTS

We have seen what can happen to an animal population when its natural enemies have been removed from the environment. But what of animals that are so large and powerful they have no natural enemies? Such animals might include the Indian and African elephant, the sperm whale, and the rhinoceros. Man, of course, is not included as a natural enemy for he may choose to protect, not prey on animals. These creatures are not preyed upon, yet some factor seems to control population growth. This is often seen when the population begins to threaten the environment.

In one instance studied during the 1960's, the elephant populations in several African National Parks grew to the point that their environment was in danger. All signs pointed to a population crash, a massive die-off of one of nature's noblest creatures. It did not happen. Instead, a number of factors apparently set into motion (not consciously, of course) by the elephants themselves cut down the population. This took place over a period of many years, and effectively reduced the strain on the environment.

Among these factors was increased age of puberty—the age at which the animals mature and can produce young. In three herds, the first healthy, the second somewhat overpopulated, and the third very overpopulated, the age of puberty was found to differ by about five years. That is, in the healthy herd it was five years younger than in the very overpopulated herd. In elephants, apparently, the greater the population stress the longer the young must wait for sexual maturity.

Another factor was the time period between the birth of calves. For females in the healthy population this was about seven years. In the heavily overpopulated herd, however, calves were born about ten years apart. A female in the first herd might produce ten calves during her lifetime. In the overpopulated herd, this number dropped to four or five.

A third factor was an increased death rate among the young. Stress due to excessive heat may have been one cause. More than likely, however, poor nutrition was the major cause. In an overpopulated area, a shortage of forage plants leads to poor nutrition among the adults and the young alike. This could have produced sickly calves with lower resistance to disease.

As strange as it may appear, certain animals not affected by predators seem to have built-in methods of population control. While we know very little indeed about these control measures in wild animals, we have learned that the same mechanisms do not seem to be present in human populations. If people are to avoid the stress of overpopulation, they must consciously impose and somehow enforce birth control.

FROM MONKEY HEAVEN
TO MONKEY HELL

The langur is a leaf-eating monkey found in India and Ceylon, among other places. Langurs normally live in troops of 25 or so animals. These creatures are of interest to us because they were the subject of three separate population and behavior studies. As we shall see, the langur's space needs seem to be very important in determining its behavior.

The first of these studies was reported by Phyllis Jay in 1963. She chose a location in central India where the monkeys are fairly scarce. In the forests where the study was made each langur troop occupied a range of about two square miles. On the average, there were less than 20 monkeys per square mile. Separate troops rarely contacted each other. Indeed, each troop tended to avoid other troops.

When contact did occur, however, Jay found that conflict did not develop. This was probably a result of the fact that there were no clear-cut boundaries between troops. In addition, there were no defended territories. While the males in each troop fit into a rather rigid rank order, there was very little in-fighting or quarreling.

Jay's study seemed to show that langur monkeys are a well-adjusted, nonaggressive lot. Her monkeys never fought, never defended territories, and in general behaved themselves.

But then in 1967 biologist Suzanne Ripley made an equally detailed study of the langur population on the island of Ceylon. These monkeys were much more crowded than those of Jay's study, although the troop size was about the same. The population density here was about 150 per square mile. Also, each of these troops on Ceylon occupied only one eighth of a square mile, as opposed to the two square mile area of Jay's monkey troops.

Behavior also differed. The Ceylon langurs actively defended territories that had very definite borders. Even more interesting, these monkey troops sought combat. Early morning challenging whoops between troops quickly led to mobilization along borders. True fighting, as well as aggressive display, often took place. Rarely, however, was there significant physical harm done during the fighting. Rather, a tremendous amount of excitement was produced. It is interesting to note that aggressive behavior did not show up within troops themselves. It was all directed to other troops. Females as well as males joined in the territorial battles.

Was this dramatic change in behavior the result of greater crowding? It would seem so, for the forest in Ceylon could have supported far more monkeys—there was no shortage of food. Again, however, we must show caution in reaching conclusions, for Jay's evidence contradicts these results.

The third study of langur population density and behavior was conducted in 1967 by Yukimara Sugiyama of Kyoto University. Sugiyama chose the langur population in the Dharwar forest of western India. He found a population density of almost 300 animals per square mile. He also found the nearest thing to the wild equivalent of Calhoun's behavioral sink.

Sugiyama made a detailed study of nine troops. He found territorial behavior, but he noted also that boundary lines were poorly defined and not well defended. Contact between troops was frequent and violent. It also differed from the type of contact shown by Jay's and Ripley's monkeys. In Sugiyama's troops, when contact was made the leaders of the two troops fought alone. For some reason the other members of the troops held back. Social organization within troops was different also. Sugiyama's troops did not show the rank order among males characteristic of the Jay and Ripley troops. In fact, most of the troops Sugiyama observed had only one adult male, rather than several. He thought this occurred because rank order among males was very weak, and vicious in-fighting drove out all but the strongest male.

The ousted males formed all-male gangs that roamed the forest much like the prober rats in Calhoun's experiment. The antisocial behavior of these gangs reached a peak during mating season. Its result was the monkey equivalent of the rat rape Calhoun observed.

Sugiyama observed mass attacks by male gangs on troops containing females in season. The invaders killed or drove off the male leader of the troop and all juvenile males. They then fought among themselves for the right to mate with the females of the troop. The females showed no regret at the loss of their leaders. Instead, they responded with intense excitement to the demands of the male gang, and mated freely with the conquerors.

This behavior, however, had unusual side effects. The young, especially infants, were neglected by their mothers. Finally, each attack was concluded by the male conquerors attacking and biting to death all of the young of the troop. The parallel between this behavior and that of the prober rats is shockingly apparent.

Thus, we see that even among primates—animals that stand erect, including man—overcrowding seems to bring about stresses that lead to abnormal and antisocial behavior. The temptation is great to conclude that these findings apply also to human populations. Unfortunately, we should not do this for the evidence is not as yet conclusive.

6
POPULATION CRASH!

The idea that the maximum amount of food available is what controls population growth—the Malthusian doctrine—was mentioned in Chapter Three. We also pointed out that Malthus was wrong. We did not, however, describe any evidence to support that statement. Fortunately, there is strong evidence to discredit this theory.

During the middle and late 1940's scientists began to suspect that animal population control depended on factors other than predators and food supply. For many years, for example, population crashes—large-scale die-offs—had been observed in a number of different animals. Lemmings, shrews, and snowshoe hares, for example, show this behavior in many different parts of the world. In each case a rapid and striking increase in population numbers is followed by what looks like mass suicide. Hundreds, thousands, and sometimes millions of animals die during a very short period of time. In none of these cases, however, did there appear to be any shortage of food. Nor did the dead animals show any signs of starvation.

THE SIKA DEER OF JAMES ISLAND

John Christian was one of the men studying this problem. He had special medical training that allowed him to look at the problem from a different point of view. This was fortunate, for Christian came up with an interesting new idea in 1950. He suggested that population changes—increases and decreases—among mammals are controlled by some internal reaction to population density. In effect, he was saying that certain physiological systems within the animal's body were capable of reacting to the stresses of overcrowding.

He thought that as the number of animals in an area increases, the stress of crowding also increases until a physiological reaction causes a die-off. But to be sure, Christian needed more evidence. He wanted to study the physiology of an animal before, during, and after a population crash. He found his animal—the Sika deer—on James Island in Chesapeake Bay.

James Island lies about a mile offshore near Cambridge, Maryland. It has a total area of about one-half a square mile, and is uninhabited. The Sika deer story begins in 1916, when four or five deer were placed on the island. The deer bred freely, for there were no predators on the island. Over the years the herd grew steadily in size. Finally, around 1955, it numbered 280 to 300 animals—a population density of about one deer per acre of land. It was at this point that Christian entered the picture.

His first step was to shoot five deer and make a detailed study of the condition of their vital organs. This task completed, Christian's team of scientists settled down to wait for the expected population crash. It came during the first three months of 1958. Over half the deer on the island died during this period; nine-tenths of the dead animals were does and fawns. Christian's group was able to recover 161 of these carcasses.

More deer then died the following year, with the population eventually leveling off at about 80 animals. Numerous additional bodies were taken for study after the first die-off and up to the spring of 1960, when the experiment ended.

Why did almost 200 deer die suddenly during a two-year period? Starvation was clearly not the reason, for the deer collected seemed to be in very good condition. Their coats were good, they had well-developed muscles, and fatty deposits were found between the muscles. Something else must have been wrong.

There are three important time periods in this experiment; (1) from 1955, when the first sample animals were taken, up to January, 1958; (2) the first die-off period during the first quarter of 1958; and (3) the second die-off period of 1959, through the time of population leveling, and into 1960. Animals taken from each of these periods tell the story of what happened.

To begin with, carcasses taken after the die-off of 1958 were the same as those taken before except for one feature. These animals were larger and heavier. There was plenty of food during 1958, but apparently the stress of overcrowding prevented the pre-die-off deer from reaching their growth potential.

Changes in the adrenal glands of the deer are even more significant. These glands play a vital role in growth, reproduction, and the body's defenses. Their size and weight vary in terms of the demands placed on them. Thus, if an animal is under heavy stress for long periods of time, the adrenals increase in activity and become enlarged. We'll say more about the adrenal gland later.

Christian found that the weight of the adrenal glands of the Sika deer remained about the same from 1955 through the die-off period of 1958. But then a sharp drop occurred in the weight of the glands. In mature animals a 46 percent decrease was noted. In immature deer the weight of the adrenals dropped 81 percent after the die-off of 1958. Changes in the cell structure of the adrenals were also noted.

Based on this evidence, Christian concluded that the stress of overcrowding brought about physiological changes that in turn led to death from shock. There was no indication of starvation, infection, or any other obvious cause of death.

MINNESOTA SNOWSHOE HARES

The snowshoe hare is an animal that suffers periodic population crashes. It does not possess any method, such as that of elephants, for regulating its numbers. Thus, its population peaks and then drops sharply in cycles. Once again, however, the die-off seems related to the stress of overcrowding.

The first attempt to explore the physiology of the snowshoe hare population crashes took place during the late 1930's. At that time a team of field biologists collected dead hares during a die-off. There was little evidence of disease, yet the animals died in a very strange manner.

Whether behaving normally, or in a torpor (a sleepy state), when a hare's time to die arrived the animal was seized by convulsions that quickly led to death. Even apparently healthy hares taken captive died almost immediately in their cages.

Normally hares get along very well in captivity. The biologists did autopsies on a number of the dead hares. They found a certain amount of breakdown in liver tissue, less than the normal amount of sugar in the blood, and minor internal bleeding. They called it death due to shock disease.

THE "BRUCE EFFECT"

As suggestive as the findings on the Sika deer and snowshoe hare are, they are not conclusive. Too many factors in the wild state are not controllable. Some might not even be known. Because of this a carefully planned laboratory experiment was needed to prove that the stress of overcrowding produces physiological changes. As often happens in science, this experiment grew out of work that was begun for a different purpose.

Two English scientists, H. M. Bruce and A. S. Parkes, were studying the reactions of birds and mammals to stimulation by unusual sights and smells. In one experiment conducted during 1966, they discovered that a newly pregnant house mouse would lose the pregnancy if she mated with any strange male mouse within four days of conception. This is the so-called "Bruce effect."

House mice are territorial animals. Thus, the female under normal population conditions does not meet any strange adult males. She might very well, however, meet strange males under crowded conditions.

Their curiosity stimulated by this interesting result, the researchers did several additional experiments. They found, for example, that it was not necessary for the pregnant female to mate with a strange male. Her pregnancy would be blocked just by the male's presence in the cage.

Finally, the scientists tried the effect of scent alone. They placed the pregnant female in a cage from which they had just removed a strange male. The female's pregnancy again failed. All she had to do was smell the presence of the strange male to lose the pregnancy. To prove that it was indeed the scent of the strange male that blocked pregnancy, the scientists destroyed the sense of smell in several pregnant females. In these females the smell of a strange male did not block pregnancy, showing clearly that the male's scent triggered a physiological change ending the pregnancy.

Another more recent experiment shows that the Bruce effect is not isolated to mice. Meadow voles were used in this study. Twenty females were mated and then allowed to remain with their mates. Of the 20, 16 produced litters. Another 20 females were then mated. These females, however, were exposed to strange males just after mating. Only five of the females produced litters.

The Bruce effect is clear evidence that crowding produces changes in the body that in turn affect reproduction—at least in some animals. Thus, this is in effect a case of natural birth control. As population density increases, the space needs of the individual animal are violated. A physiological reaction then relieves the stress of high population density by reducing the birthrate. There is no evidence of anything like this effect in man.

LEMMINGS—THE "FOUR-LEGGED MOWERS"

The periodic mass die-offs of the lemming—a tiny Arctic rodent—are well known in both fact and legend. Every three or four years lemmings appear in large numbers. Gradually increasing their breeding rate, they finally multiply to a population density that they apparently cannot live with. At this point the die-off occurs.

Some of the millions destined to die simply disappear; these animals are probably eaten by predators. Others begin a mass migration in all directions. But there is no order to their movements. Many fall off cliffs to their deaths. Others drown attempting to cross bodies of water lying in their way. Still others march on until they disappear as if by magic.

A lemming horde on the march is an awesome sight. A lemming eats its own weight in food every 24 hours. Thus, an army of the animals on the move brings great destruction to farmlands. First, all of the greenery is eaten. When this is gone, the animals attack the root stocks. Scandinavian farmers, observing this destructive behavior, nicknamed lemmings "four-legged mowers."

Why this strange cyclic behavior from a normally innocent tiny animal? No one really knows the answer, although some facts have been established. Most scientists agree, for example, that overcrowding is what triggers the die-offs. Some believe that the adrenal gland is involved, just as in the Sika deer of James Island. The theory is that under the stresses of overcrowding the adrenal glands become overactive. These glands then flood the animals' bodies with adrenaline, a chemical substance that increases the rate of the body's physiological functions. The overworked glands and body systems break down and the animals die.

Another theory is based on the discovery of a new substance in lemming blood. This substance seems to act as a sort of antifreeze during the very cold Arctic winter. It increases blood flow to the skin and legs, for example, permitting the lemmings to be active during severe cold. They forage about for food beneath the snow all winter long.

It has been suggested that the warm weather of spring and summer changes the antifreeze into a poison that attacks the nervous system. This may be so, although it does not explain those years when no die-off occurs. If this substance is involved in the die-offs, it is more likely that it is altered chemically by physiological changes brought about by the pressures of overcrowding.

In any event, it is clear that some animals, such as the African elephant, have built-in physiological methods for controlling their population levels. Other animals do not. In these, the lemmings,

Scientists do not as yet fully understand what causes Arc

voles, and Sika deer, the pattern is often one of rapid population growth, overcrowding, and then massive die-off or population crash.

Scientists agree that man has no built-in physiological method for controlling his population growth. Thus, the current world-wide human population explosion threatens in at least two ways. It is quite possible that man will outgrow the ability of the planet to support his numbers. It is also possible, although not proved, that extreme overcrowding will produce physiological changes that in turn will lead to a human population crash.

ʒion lemmings to experience massive die-offs.

7
THE
BIOLOGY
OF
STRESS

The experimental results with the Sika deer, snowshoe hare, meadow vole, and other mammals demonstrate that when animals are severely overcrowded something happens to their physiology. Often the overcrowded creatures are smaller in size than those not crowded. Almost always the adrenal glands are enlarged. Other tissues also frequently show physical changes. Sometimes, as in the house mouse, just the smell of a strange male can so affect a pregnant female that the pregnancy is ended.

Biologists usually think of man's adrenal glands in terms of emergency reactions to danger. Sudden disease, broken bones, burns, and the contact of a game such as football all start the adrenal glands pumping. This isn't all, however. Fear, anxiety, the shock of a loud noise, even being in a strange and unknown place will do the trick.

In recent years it has been discovered that over-crowding has the same effect. The adrenals react by increasing in activity and becoming enlarged. This indicates that the animal—whether man or beast—is under severe stress. The body is reacting to the stress the only way it can—by increasing the animal's awareness, alertness, and readiness to defend itself. But this takes energy, and burns up sugar. It also traps the animal's body in a vicious physiological circle that can lead to death from shock disease.

THE PITUITARY-ADRENAL SYSTEM

To fully understand what shock disease is, it is necessary to look a bit closer at the body's defensive reaction system. This involves the pituitary gland (a small but vitally important organ located deep within the brain) as well as the adrenal glands. These are attached to the kidneys, which are located in the lower back region in mammals. Both the pituitary and adrenal glands secrete hormones (sometimes called chemical messengers) directly into the blood stream. The hormones then stimulate other organs into action.

When an animal experiences stress, the central nervous system delivers "news" of the stress to the brain. This stimulates the release of a succession of hormones, which in turn eventually stimulate the adrenal glands. At this point the adrenal glands release the mixture of hormones that put the body in "fighting" shape.

We mentioned earlier that it takes energy for the adrenal glands to put the body on the alert. This energy comes from sugar, which is stored in the liver. If we think of the liver as a bank, the production of shock disease by stress is very similar to overdrawing one's bank account. In effect, the reaction of the body is to attempt to take more sugar than there is available from the liver.

When stress persists for a long time, physiological changes take place. The fertility of the stressed animals may decrease, as in elephants. Vital organs may undergo changes, as in the Sika deer on James Island and the dead snowshoe hares in Minnesota. Changes in behavior may also take place, as in lemmings.

But when things become this bad, just a tiny increase in stress can overdraw the bank account. The adrenal glands react by sending a jolt of hormones into the bloodstream. This, however, uses up the last of the sugar available, and the brain is suddenly starved for sugar. This condition is called *hypoglycemic shock.* It is probably what occurs in the population crashes, in the massive die-offs of severely overcrowded animals.

CROWDING AND SHOCK DISEASE

His interest aroused by the Sika deer results, John Christian later made a study of woodchucks. He wanted to know how the adrenal glands changed from season to season as the activities of the animals changed.

This study lasted four years. In all, nearly 900 animals were collected. The adrenal glands of these animals were carefully weighed, and the weights were entered into a table showing the season when the animal was taken.

Christian found that the weight of the adrenals increased sharply twice each year. The first increase took place between March and June—the woodchuck's mating season. During this period males were competing strongly for mates. They were also active for much longer time periods each day. But most important, more woodchucks were concentrated in a given area than at any other time of the year; they were more crowded than usual.

The weight of the woodchucks' adrenals then dropped in July, when both aggressiveness and crowding eased up somewhat. The weight then jumped again in August, when the young were moving out to establish their own territories. During this period, there was much movement and frequent conflict among the animals. Christian also found that when male woodchucks were most aggressive, they tended to stay apart from other woodchucks. Apparently they needed more space.

Now, how do these results tie in with death from shock disease? To begin with, we have seen that when animals are crowded they become more aggressive. But the woodchuck experiment, as well as others, tells us that as animals become more aggressive they need more space. The real trouble then starts when there is no more space, when the animal showing aggression cannot find the space needed to relieve its tension.

This occurs when a population is severely overcrowded. The stress of overcrowding usually starts a chain reaction of behavioral and physiological reactions. Excessive aggressiveness, heightened sexual activity, and other strange forms of behavior (the behavioral sink) explode within the population. But this puts tremendous pressures on the animals' adrenal glands, which respond by using up the last of the available sugar. Hypoglycemic shock is the result. The massive die-off of a population crash then follows. Shock disease is likely the major cause of the die-offs described here.

CROWDING AND MAN'S ADRENALS

Do these results apply to man? This is a difficult question to answer, for clearly we cannot deliberately overcrowd human beings and then sacrifice them in order to study the condition of their organs. It might be possible to run blood tests for adrenal hormones under a variety of conditions, but this too would be difficult. Not many people would volunteer to live the way Calhoun's rats lived. Unfortunately, to get reliable results, such living conditions would probably be required.

Thus, to find any evidence that might apply to man it is necessary to look at research in other areas. The work of Robert Henkin of the National Heart Institute is a good example. Henkin studied the connection between adrenal hormones and sensory perception. His findings are very interesting, and may tell us something about man's fate in a severely overcrowded world.

Henkin first studied patients whose adrenal glands had been removed or were functioning poorly. He found that these patients, who lacked a normal amount of adrenal hormone in the blood, showed a marked increase in sensory perception. Their senses of taste, smell, and hearing were greatly sharpened and intensified.

When Henkin studied patients whose secretion of adrenal hormones was above normal, however, he found that the senses were significantly dulled. In other words, when the blood carried more than the normal amount of adrenal hormone, the patients' ability to detect tastes, odors, and sounds was greatly decreased.

Finally, Henkin studied this effect in normal people. He discovered first that the adrenal hormones are secreted in a daily cycle. He then discovered that the intensity of the senses varies with the amount of hormone in the blood. It was not due to some other factor connected with the illnesses of the patients.

Think a moment about what this may mean. We know that severe overcrowding produces stresses. This in turn may cause the body to suffer harmful physiological changes. In particular, when under stress the adrenal glands secrete hormones without let-up. More than the normal amount of adrenal hormone is present in the blood stream at all times.

But Henkin found that in human beings the senses are dulled when too much adrenal hormone has been secreted. It follows that people suffering from severe overcrowding will probably react sluggishly to what is going on around them. They will taste less, smell less, and hear less. If you've ever had your hearing or your senses of taste and smell reduced by a bad head cold, you have some idea of what this would be like.

People whose senses do not respond fully to the outside world tend to become withdrawn. Man

*Sleepy? Or feeling the effects of stress
due to overcrowding?*

needs sensory stimulation to function at his best.
If his senses are damaged or impaired for long
periods of time, it is possible that he will not be
able to work, think, or play normally. This descrip-
tion brings to mind once again the "zombies," the
poor folk of the imaginary city described at the
beginning of the book. These people, as you recall,
collapsed completely under the stresses of their
overcrowded city.

8
THE
EVIDENCE
ON
MAN

So far in this book we have talked primarily about how severe over-crowding affects animals. It was suggested also that these effects might occur in man too, if his numbers continue to grow unchecked. It is important to point out again, however, that there is very little real evidence on the effects of overcrowding in human populations. Most, if not all, of what the various experts are saying is based on *extrapolation.* That is, they suggest that the experimental results obtained using animals may also apply to man's population problems and how man may be affected by severe overpopulation.

This is clearly not good enough. If we are to survive the coming population squeeze, we must learn how to make the correct decisions about population growth. Unfortunately, this isn't easy, for man is not an easy animal to experiment with.

FIRST EXPERIMENTS ON MAN

Despite these difficulties, however, a modest start has been made. We say modest because the experiments conducted to date have been very simple. Yet they have yielded important results, and show the way to further experiments. Paul Ehrlich and Jonathan Freedman got things started during the late 1960's at Stanford University. These scientists ran a series of tests that all had the same basic structure. Groups of people spent a certain amount of time in rooms that were either very crowded or not at all crowded. The difference between crowding and noncrowding was quite simply arranged. The same number of people were put either in a very small room or a large room. For example, each person might have less than four square feet of space in a crowded room, but as much as 20 square feet in a noncrowded room.

The first experiments attempted to show if crowding had any effect on ability to carry out simple tasks. The experimental subjects memorized words, solved puzzles, formed words from scrambled letters, counted clicks coming over a loudspeaker, and so on, in both small and large rooms. The results? There was no difference in performance between the crowded and noncrowded groups. Nor did any effect show up as more time was spent in the rooms. Apparently crowding has no effect on man's ability to perform simple routine tasks, at least to the extent of this experiment.

The second set of experiments, however, produced quite different results. These studies show that population density has a decided effect on complex types of behavior that involve emotional reactions. In brief, competitiveness, vindictiveness, and affection toward others are affected. Even more interesting, men and women seem to react in opposite ways to overcrowded conditions.

In one study, groups of high school students met together in both small and large rooms. The students first got to know each other, and then moved on to more difficult tasks. They discussed various controversial topics, and ended up playing a game that sets cooperative behavior against competitive behavior.

The rules of the game were quite simple. There were four players. One trial consisted of each player secretly writing either of the colors "red" or "blue" on a piece of paper. The players knew in advance, however, that if all four chose blue, they would each be rewarded with a five cent bonus—this represents cooperative behavior. On the other hand, if one player chose red while the other three chose blue, the player choosing red would be rewarded with 30 cents while the others would lose

20 cents apiece. Choosing red in opposition to choosing blue is competitive behavior. If more than one chose red, however, then all players would lose 20 cents.

Thus, the cooperative approach (all players choosing blue) produces a small but sure reward. With many trials, this could add up to a nice little bundle of cash. The competitive approach, however, offers a much larger reward per trial and the opportunity to build up winnings much faster. The risk of losing, however, is much greater and must be considered when the choice is made.

The results of this experiment are unusual and interesting. A definite pattern of behavior was detected. The boys played the game more competitively in the small room than in the large room. The girls, on the other hand, were more competitive in the large room than they were in the small room. In terms of crowding, this suggests that males are more competitive when crowded and females are less competitive when crowded.

Another experiment used a different set of people and probed even more complex behaviors. For this experiment men and women were recruited by newspaper classified ads. The people chosen came from a wide variety of backgrounds; they also represented a variety of educational and income levels.

Groups of these people also were placed in either small or large rooms. They too began the experiment by getting to know each other by taking part in informal discussions. They then listened to tape recordings of courtroom cases, and served as the "jury" for each case. On all cases each "juror" recorded his vote and his sentence independently and privately. In addition, each person answered a questionnaire that asked (1) how much he or she liked the other members of the group, and (2) how much he or she enjoyed the experimental sessions. The group also answered other questions designed to probe the individual's emotional reactions to the experience.

The results of this experiment were similar to those of the "red vs. blue" experiment. There was some difference in emphasis, however. In this case the differences in behavior between the small and the large room were much greater for the women than for the men. When in the small room the men were more severe and the women more lenient in judging the courtroom cases.

The answers to the questionnaire show that crowding has an even stronger effect on how people react to the experience. The men, when crowded in the small room, found the experience very unpleasant. Moreover, they liked their fellows less and found them less friendly. Finally, the crowded men thought their groups made a poorer jury than the men working in the larger room.

For the women, every reaction was just the opposite. When in the small room, they found the experience more pleasant, and their companions more likeable. They also regarded the group in the small room a better jury than the group in the large room.

Even more interesting, when this experiment was repeated using mixed groups of men and women, no crowding effects were observed. Mixing the sexes seems to cancel out the opposite effects noted when the men and women were isolated.

Why does crowding seem to affect men more severely than women? There is no easy answer to this question, for human beings are complex creatures whose behavior is difficult to analyze and explain. Moreover, research on this problem is just getting started. Thus, the work of Ehrlich and Freedman is no more than a tantalizing beginning. Nevertheless, some experts suggest that feelings of territoriality in males explain the behavior differences. Others think that women grow up with good feelings about close physical contact with others, and therefore feel more at home when they are crowded. These theories are possibilities. More than likely, however, the real answer awaits discovery.

In another study, reported early in 1972, social scientists Omer Galle, Walter Gove, and J. Miller McPherson of Vanderbilt University measured how high population density affects human behavior in a large city. The city was Chicago. The results of the experiment support, but do not prove, the idea that overcrowding has a negative effect on certain human behavior patterns.

The behaviors studied were (a) *mortality*—the death rate of a given overcrowded area compared to the death rate of the city of Chicago as a whole; (b) *fertility*—the numbers of births per 1000 women between the ages of 15 and 44; (c) *poor care of the young*—the number of children under 18 years of age receiving welfare or some other form of public assistance; (d) *aggressive behavior*—the number of male youngsters brought to court on delinquency charges; and (e) *withdrawal or mental collapse*—the number of admissions to mental hospitals per 100,000 persons in the community.

Now, what did Galle, Gove, and McPherson find? Their first task was to find instances of population density that reveal how the behaviors above were affected. For mortality, fertility, poor care of the young, and aggressiveness they found that the average number of persons living in a room is the most important indicator. They were unable to find a reliable indicator for mental withdrawal, however. Let's look at each of the behaviors in turn to see the possible effects of overcrowding on man.

Mortality. Deaths appear to be higher than normally expected when the number of persons living in a single room is high. In some slums, for example, as many as eight to ten or more people often live in one room. The higher death rate under these conditions seems to be the result of greater exposure to infection, constant weariness and irritation caused by disturbance by the activities of others, and the withdrawal probably associated with overcrowding.

Fertility. In most animal populations, overcrowding leads to a drop in the number of young born. The opposite appears to happen with human beings—the greater the number of people crowded together, the more children born. Remember, however, that animals compete for mates and must acquire territories during each breeding season before successful breeding can take place. This is not the case with man, who is capable of breeding throughout the year. In addition, people who are overcrowded seem less able to understand how more and more children will affect their futures. Thus, they often fail to take steps to control the number of children born.

Poor care of the young. Because overcrowding produces tensions, irritations, mental withdrawal, weariness, and poor health, care of the young suffers. As a result, a higher level of welfare assistance to care for neglected children was found among overcrowded families.

Aggressive behavior. Because parents under overcrowded conditions are likely to be tense, irritated, weary, and withdrawn, it follows that children will find such homes unattractive. These children spend as much time as they can out of the home and away from parental control. The result is an increase in the number of juvenile gangs, along with more frequent clashes with the law and an increase in the number of youths appearing before juvenile courts.

Mental withdrawal. Galle, Gove, and McPherson were unable to find any connection between overcrowding and admissions to mental hospitals. They point out, however, that they were probably not measuring the correct indicator for mental withdrawal. They did find that more people living alone are admitted to mental hospitals. They go on to suggest that overcrowding may have originally triggered the mental disorders that led to living alone. They were unable, however, to find support for this idea within the framework of their study.

RANK AND STRESS IN MAN

When faced with the limited amount of direct experimental evidence on crowding in human populations, indirect evidence must again be used. As pointed out earlier, conclusions drawn from this type of approach must be taken with a grain of salt. Nevertheless, there is often something worthwhile in a new look at experimental results.

Most animals exhibit a "pecking order." That is, among groups of animals of the same species there is a status structure, with high ranking individuals and low ranking individuals. We know that the low ranking individuals in many animal populations sometimes sicken and die for no apparent reason. Also, some of the experiments described in this book have produced animals whose behavior is extremely unusual. These animals, for example the drop-outs, the probers, and the loners of Calhoun's rat experiment, must be considered sick when their behavior is compared to that of normal animals. There is evidence also that low ranking animals suffer from enlargement of the adrenal gland. But this, of course, is precisely what happens to animals when severely overcrowded.

Thus, there seems to be a physiological similarity between the stress of being "low man on the totem pole" and the stress of overcrowding. As a result, a study of the effects of rank in men should tell us something about how man will react to severe overcrowding.

During the late 1960's the results of a medical study on 270,000 male employees of the Bell System were announced. This mammoth investigation tied together job achievement, educational background, and the incidence of heart disease. The results were startling, to say the least. Most Americans, for example, firmly believe that the price of great success is often broken health. Ulcers, nervous breakdowns, heart disease—these are the afflictions that haunt the occupants of the executive suite. The Bell study, however, indicates that just the opposite is probably the case.

Let's look at the results. Starting at the bottom of the company's pecking order, we find that ordinary workmen suffer heart attacks at the rate of 4.33 per thousand men per year. At the next step up, the foreman level, there is a slight increase to 4.52 heart attacks per thousand per year. The count then drops to 3.91 for supervisors and local area managers. At the next level there is an even greater drop. For general area managers the rate is 2.85. This brings us to the senior management level, to the high achievers of the executive suite. Surprisingly, and quite contrary to the popular myth, heart attacks occur in this group at the low rate of 1.85 per thousand per year.

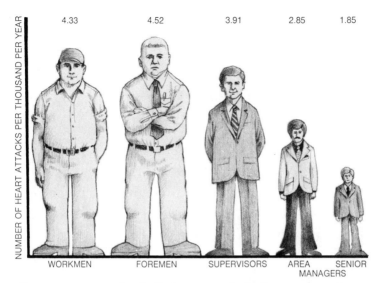

NUMBER OF HEART ATTACKS PER THOUSAND PER YEAR

| 4.33 | 4.52 | 3.91 | 2.85 | 1.85 |

| WORKMEN | FOREMEN | SUPERVISORS | AREA MANAGERS | SENIOR |

Contrary to popular belief, one study indicates that the healthiest place to work is apparently the executive suite.

This indicates that the higher a man's rank the lower the chance that he will have a heart attack (at least if he works for the Bell System). The opposite is true also. As a man's rank decreases, the possibility that he will suffer a heart attack increases. Now let's take this idea one step further. Let's suppose that the stresses that affect man low in the pecking order are similar to those of overcrowding. If this is so, then it is conceivable that the more overcrowded we become the greater the possibility that we as individuals might suffer from stress disease.

ONE VIEW OF THE FUTURE

Jay Forrester, a professor at Massachusetts Institute of Technology, is interested in how human social systems work and in how men attack the problems that beset humanity. But since very few of man's present problems are completely unrelated to the coming population squeeze, Forrester's work has something to say about overcrowding also.

To begin with, Forrester feels that man's inability to solve his major problems is the result of a sort of genetic nearsightedness. His studies lead him to believe that the human mind is not adapted to understanding how social systems really work; that instead, man quite naturally looks only in the immediate past for the cause of a problem. It is thin ice that causes the boy to fall into the lake, not the curiosity that made him venture out onto the ice in the first place.

Forrester cites many examples to support this idea. For example, when automobile traffic fills the highways in one long and endless traffic jam, our answer is to widen the roads and/or build more highways. But this merely attracts more

cars and more traffic and soon the highways are jammed once again. In another example, what happens when a city's rapid transit system, its buses and subways, are in financial trouble? Almost always fares are raised to bring in more money. But this has the effect of persuading more people to use their cars. As a result, the highways become more clogged and the transit system continues to lose money. Earlier in the book we spoke of the high-rise apartments put up to provide new housing for people living in leaky, crumbling, rat-infested slums. The planners of these projects, even though they are trying to do the right thing, often fail to consider basic needs of the people. For example, the old neighborhoods are destroyed, and the people are thrown in with strangers. Unrest usually follows.

Moreover, in many cases these developments lead to even worse slum conditions. Instead of relieving overcrowding, an even more dense population results because more apartments are squeezed into the same area. The low income families flock to the new buildings, only to discover that there are too few jobs and that the region cannot support them. How could it, when previously it could not support a smaller population?

The Pruitt-Igoe project in St. Louis is a perfect example of the failure of low-cost inner-city housing projects. Twenty-six 11-story glass-and-concrete apartment buildings went up 15 years ago. Today some are boarded up, and others are being torn down. The unrest of the people led to such vandalism and physical decay it was impossible to keep the buildings in good condition. The poor, even the poorest of the poor, refused to live there. Today the remaining buildings stand vacant, mute testimony to the wrong solution to the problem.

Forrester has extended his theory to deal with the social system of the entire world. This included many assumptions, and very elaborate computer calculations, far more than men working by hand could hope to complete in many, many years. In one of his projections, Forrester assumed that we would not run out of our valuable natural resources prior to the year 2100. That instead, man will find substitutes for metals, and power sources that will reduce the demand on coal, gas, and oil reserves.

Let's see what might happen, assuming Forrester's theory is correct. Man, with his love of growth, spurred on by pressure from the "have-not" nations who want to "have" just as we do, will continue to build and industrialize. But pollution too will grow as industrialization expands. Couple these effects with continued population growth, and disaster may occur shortly after the year 2030.

World population may at this point have reached the nine billion level. If this happens, the quality of life may start to drop alarmingly.

Then, according to Forrester's model, the earth will suffer a human population crash. Man's numbers will decrease by about five billion, a die-off almost impossible for the human mind to grasp. If you think it isn't likely to happen, consider this. The only path to the so-called life of plenty, to the standard of living enjoyed in the West, is industrialization. Do we say to the poor, to the underprivileged of the world, "Hold it, the world can't take any more. You must stay poor. You must remain a have-not"? Probably not. But if industrialization continues, and the standard of living for the rest of the world is raised to the level enjoyed by most Americans, the pollution load on the environment could be ten times today's level.

The population crash will be caused in part, at least, by the stresses of overcrowding. Crime, war, and disease would take an even greater toll than they do today, for people will find themselves too close to each other for peace and comfort. Starvation will take uncountable millions. In some parts of the world, people will kill each other for standing room. In other parts of the world, people will be forced to live in places too dangerous for man because there will be no other place to go. Many of these people will die suddenly from the effects of natural disasters. Some experts, for example, believe that overcrowding was the main cause of the hundreds of thousands of deaths in the cyclones that hit The Bay of Bengal in 1970 and 1971. Hordes of people, having no other place to live, settled on the Ganges River delta. They had no choice. The region, however, is so exposed to storms and flooding no one would live there if they could settle in safer areas.

Forrester's studies are very disturbing. They are extremely speculative also. It should be remembered when reacting to his predictions that they are the result of little real evidence and much scientific guesswork. Despite this fact, however, Forrester may be telling us what it could be like in the year 2100. He predicts disaster for mankind if industrialization continues to spread and if the population explosion continues unchecked. There seems to be only one way to ultimately provide the standard of living of the West for the entire world. This is to hold both world population and industrialization at levels lower than current averages. As drastic as this sounds, *if* Forrester's projections are accurate, man's only hope lies in *reducing* both world population and the level of industrialization. The opposite—growth, with its ever increasing pollution and population—is unthinkable.

9
OF MICE AND MEN

John Calhoun, the inventor of the term "behavioral sink," is far more than just a behavioral scientist. At times during his professional life he has been a psychologist, a mathematician, a sociologist, an ecologist, and a philosopher. Today he is deeply involved in human ecology—the study of the relationship of man to his environment. Calhoun's special interest is what may happen to man if human population grows out of sight. In this work, he is but one of a rather small group of scientists with similar interests. His research, however, is particularly intriguing because he has made an effort to use its results to predict what the future will be like. Because of this work, he has been called both a "prophet of doom" and a "prophet of hope."

This chapter describes Calhoun's current work and his ideas about the future. After reading it, you decide. Is he a prophet of doom or a prophet of hope?

Intrigued by the findings of his rat experiment, which were described in Chapter Three, Calhoun later began a large-scale population study using mice. This experiment is not scheduled to be completed until 1973. As mentioned, however, Calhoun has collected enough data to formulate a philosophy of human population for the future. Like Forrester, he predicts a massive population crash if we make the wrong decisions about the earth's future population. Unlike Forrester, however, he shows a possible way for continued growth of the human potential.

Calhoun suggests that we have just entered a 50-year crisis period. Man must decide during this period what his fate is to be. If his decision is the correct one, Calhoun calls the end of the 50-year period "dawnsday." If this all important decision is the wrong one, the end of the decision period becomes "doomsday."

A WORLD IN MICROCOSM

Calhoun admits freely that mice do not behave the same way men do. He agrees too that it is difficult, and perhaps inaccurate, to apply his mouse-population findings to men. Nevertheless, he sees enough of value in the mouse experiment to use it as the basis for his population philosophy of the future.

The experiment began with the introduction of four pairs of mice into a nine foot by nine foot mouse "universe." This world in miniature contained enough food, water, and shelter to support a population of 3,000 mice. There were 256 tiny apartments, numerous feeding bins, and a central floor area. As the mouse population grew, the feeding bins and the central floor area became the social centers for the mouse universe.

The original eight mice, after a short period of adjustment, quickly multiplied to about 150 animals. This number, Calhoun now maintains, is the ideal population for the experimental set up. With 150 animals in the pen each mouse would experience just the right amount of social contact with other mice. Thus, each mouse would receive its fair share of social satisfaction. Under these conditions the mice would live their lives normally, with no unusual stresses to disturb them.

This mouse world, however, differed in one very important way from a natural mouse environment. All of those factors—hunger, disease, bad weather, predatory animals, and migration—that keep population growth down in the wild state were eliminated. The mice in Calhoun's miniature

John Calhoun in his mouse "universe." Note water bottles (around the top), nesting apartments (around walls below bottles), food containers (on walls between apartments), and cans of nesting materials (off the floor).

world could die of only one thing—old age (unless something unforeseen struck them down first).

As a result, the population increased rapidly to about 620 individuals. The rate of growth then decreased, but the population grew steadily until it reached some 2,200 animals. This point was reached in a bit more than two years. At this time the mice stopped reproducing. In over a year not a single mouse was born. Some 600 mice, however, died of old age, reducing the population to 1,600.

Calhoun then discovered that the females remaining in the population had lost their ability to reproduce. They either were too old to bear young, or the stress of overcrowding had made it impossible for them to mate and conceive. Thus, after about a year of die-off it became clear that the remaining population was doomed. Even though the experiment has not been completed, Calhoun feels that there is no turning back for this population. It has "overlived" itself, and will gradually decline in numbers until the last mouse dies.

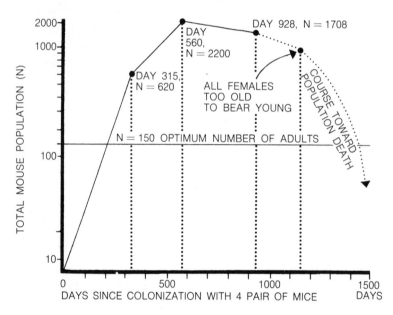

The life history of the experimental mouse population in Calhoun's mouse "universe." These mice faced no dangers, except one—overcrowding.

ANOTHER BEHAVIORAL SINK

Shortly after reaching the 150 population level, the mice formed 14 social groups of about 10 mice each. These 14 groups filled all of the social space available within the experimental pen. For this strain of mice, there was no room for any more members in any single social group, nor was there room for any more groups of ten. In other words, the pen was only large enough to support 14 social groupings containing a total of about 150 mice. All mice born after this social structure was established had but two choices—they spent their lives outside the social organization of the colony, or rarely, a young mouse succeeded in replacing a mouse that had gotten old or died within a group. Since only a few mice within the social groups died, the vast majority of the 2,200 mice at the peak population level lived outside the social fabric of the colony. This had a disastrous effect on these mice.

Unable to leave the pen and unable to gain entrance to one of the social groups, the excess mice withdrew into a huddled mass in the middle of the pen. After repeated rejections by the "organized" mice, these outsiders sharply reduced their movements—apparently to avoid being noticed. They simply huddled together or rested motionless alone. Within the huddled mass, however, there were frequent violent outbursts among the rejected mice. Moreover, because these mice were so withdrawn, they did not flee when attacked. Trembling violently, they remained in place and suffered severe bite wounds.

The rejected mice learned to carry out only the simplest of behaviors, such as eating and drinking. They never learned to fight to achieve rank, or mate as normal mice do. Thus they were denied the activities of socially organized mice. They matured into passive blobs of living tissue, sleek and fat physically, but totally unable to take part in the normal activities of a mouse. As Calhoun comments, these animals were no longer mice, "they were just living fossils."

There could be a message for man in the experience of Calhoun's rejected mice. For example, the violence in our slums in recent years may correspond to the violent outbursts among the rejected mice. Taken this way, the violence can be viewed as a warning that something has gone wrong within the population.

It is also interesting to consider whether the rejected mice can learn normal behavior patterns. Some, therefore, will form the basis of another Calhoun experiment. These mice will be placed in a completely normal mouse environment to see if they can recover from the rejected state. Calhoun suspects that they cannot and further suggests that this failure may apply also to human beings. If this is so, it would probably do no good to remove a child from the slums and place him in an ideal environment—mainly because it would be impossible to repair the damage done during the child's earliest years.

The results of the experiment further suggest that severe overcrowding affects far more than the simple business of eating, drinking, and finding shelter. The rejected mice, for example, succeeded only in staying alive. They did not succeed in acquiring the behavior characteristics of normal mice. Thus, behaviors related to aggression, territoriality, status formation, courtship, mating, and raising the young were denied them.

If overcrowding becomes too severe, the same sort of problems will probably occur with people. Long before it will be impossible for man to feed and house himself, for example, his ability to think and act creatively will be blocked. The production of new ideas will stop, but in addition, man's capacity to use ideas developed earlier will greatly decrease. In short, what Calhoun calls the *maximum idea carrying capacity* of all mankind will have been reached. A day might come, for example, when the human mind will no longer be able to invent a better mouse trap (no joke intended). What is worse, at the same time, the human mind may lose its ability to use the mouse traps already in existence.

This possibility is neither attractive nor popular. Many experts, in fact, disagree completely with Calhoun's theories. His reaction is simple and straightforward. He bows to his critics, but insists that what he has discovered must be followed up, for the very survival of mankind may be at stake.

Calhoun is not the type of man to stop at a prediction of doom for mankind. He has developed a philosophy of human population that may very well show the way for man's continued evolution and development. It is this work that sets Calhoun apart from his scientific colleagues, and makes his thinking important for all of mankind.

THE STAGES OF HUMAN EVOLUTION

As an outgrowth of his interest in human ecology, Calhoun has proposed a new approach to human evolution. He suggests that man as we now know him is the result of two very important evolutionary events. The first of these events occurred about 100 million years ago.

Man was a different animal at the time of this change. He was not *Homo sapiens,* but something much more primitive. We don't know precisely how this "preman" looked and acted, or how he made his living. We do know, however, that this creature discovered the many advantages of living together in small and compact social groups. The discovery, or invention, of the social group may seem at first glance an insignificant event. A measure of its importance, however, is how far men have carried the business of banding together in groups. Today, for example, it is possible to divide the world into just two groups—the anti-Communist West and its allies, and the Communist East and its allies.

But according to Calhoun, the most important outcome of the invention of the social group was the evolution of two types of "brains." One, of course, was the development of the physical brain, the organ that now lies beneath the skulls of all human beings. The other, completely dependent upon group social action, was the evolution of a group type of thinking—a collective "brain." Instead of living only in terms of the limits of one individual's mental and physical ability, preman learned to relate to a group. He evolved the ability to cooperate with his fellows, and in so doing vastly increased the creative potential of his kind.

This expansion of his potential is the key to what then happened. With this new capacity to discover, store, and process information, preman evolved gradually to the creature that populated the earth some 50,000 years ago. Physically, he probably looked very much like 20th Century man. Creatively and intellectually, however, he still had a long way to go.

At the point in time that we are discussing, all the habitable land available to preman was fully occupied because he lacked the technological ability to live anywhere on earth except in the tropics and subtropics. Thus, he was confined to a rather narrow habitat on either side of the equator.

The population of the world at this time was perhaps 4½ million. According to Calhoun, the evolution of man would have stopped at this point if the second major evolutionary event had not occurred. Preman 50,000 years ago had gone as far as he could go. He had made the first giant step toward a community life. Because of his rather large space needs, however, he had to limit contact with his fellow creatures. Only by staying on the move, killing, or being killed, could he limit these contacts and survive as a species. At a population of 4½ million, the earth was fully populated.

Put another way, this creature was a slave to his environment. Since he could not extend his geographic range, the quality and the potential of his life were severely limited.

Thus, despite the advantages of a spoken language and the beginnings of social organization, preman was at an evolutionary dead end. Without a drastic change of some sort there was no way for him to meet and overcome new challenges, no new avenues for expansion.

But then, under the twin pressures of increasing numbers and increasing physical contact, preman evolved the ability to use *conceptual space.* Conceptual space is perhaps better expressed as *the world of ideas.* This was all preman needed to surge forward in a new burst of evolutionary activity.

In other words, conceptual space is a pool of information. It is all of the knowledge of the universe. Man, for he truly became man with this event, could now discover more information. He could then use this additional knowledge to create new rules, codes, and theories to help him cope more effectively with the physical and social environments.

We know that man was successful in this new venture because human population has soared in the past 40,000 years. During this growth period, the time it has taken world population to double has been cut in half with each doubling. It took 160 years for the world's population to reach 1¼ billion by 1868; 80 years for it to reach 2½ billion by

1948; and it will take about 40 years for it to reach 5 billion.

But how does the ability to use conceptual space free man from his environment? To get at this answer, we must imagine what life was like when preman was a hunter and a gatherer.

He gathered edible plants and killed those animals he could for food. He lived in small bands probably consisting of about 12 adults and their children. It is believed that each of these bands occupied a territory about 15 miles in diameter. Each one needed the full area it occupied for food, and perhaps even more important, to satisfy the space needs of each individual within a band. Preman, like modern man, had to maintain a certain distance between himself and his fellows for peace of mind. In addition, as pointed out, all of the habitable space on the earth's surface was filled.

As a result, preman had to free himself from the environmental trap that held him. He did this by using conceptual space, and gradually created modern civilization in all of its complexity. Exploration of these new horizons, however, allowed the population to increase without threatening the individual's personal space needs. In other words, each individual man's physical space needs decreased as his ability to think, create, and use conceptual space increased.

As Calhoun sees it, man's growth is also connected to another aspect of human development. Simply put, the ease with which man can travel to various areas of the world is responsible for his involvement with people from every land.

From bands of 12 or so man's social evolution has progressed to tribes, nations, empires, leagues, and so on. But while this was taking place, the use of conceptual space broadened and man's need to communicate with others increased also.

Today the earth is peopled by nations, but there is a clear trend toward multination alliances—for example, NATO, SEATO, and the Communist Bloc. Some predict that in less than 100 years the earth will consist of but one community with common goals and a common future. At this time the world's population may very well have increased to nine billion, a number that most authorities view with great concern.

SENESCENCE, STAGNATION, OR RxEVOLUTION?

Today, man shows more concern than ever before for the fate of the world. He has come a long way. From 100 million years ago to 43,000 years ago, preman was involved in those evolutionary processes that ended with the emergence of "man." This Calhoun calls the STRIVE period. The following period, the EXPLOIT epoch, is the time span following discovery of how to use conceptual space. This is the period during which each doubling of the population required half the time of the previous doubling.

We have just entered the DECIDE period, which Calhoun believes will end around the year 2020. Some years prior to this date, however, the world's population could reach nine billion. According to Calhoun, this would be the maximum population the earth can support and still maintain the creative potential of each individual.

If so, man must determine his future prior to the end of the DECIDE period—that is, prior to the year 2020. He alone can determine whether this is to be DOOMSDAY or DAWNSDAY. He has three choices: *Senescence*—the population explosion continues; *Stagnation*—population growth is stopped but at a level close to nine billion people; *RxEvolution*—population is reduced gradually. His choice made, man enters the COPE period; here he adjusts to the future course he has selected. Let's take each of these choices and see what could happen if man were to choose one.

Senescence. This is the route of unchecked population growth. Calhoun, as well as Forrester and others, fears that this will lead to a human die-off of unbelievable proportions. Calhoun goes further,

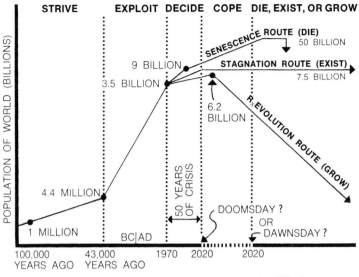

ADAPTED FROM CALHOUN

Man's population history as Calhoun sees it. We have just entered the DECIDE period. Will our choice for mankind be to die, exist, or grow?

however. He feels that as population increases beyond the nine billion mark, the potential of the individual must decrease. Thus, individual man will become less and less aware of what is going on around him as the population grows beyond the nine billion mark. And, just as Calhoun's overcrowded mice failed to learn complex mouse behaviors, man will fail to reach his potential. The production of new ideas will stop, and the ability to make use of older ideas will fall off.

As a result, man will be less able to cope with problems arising among his fellow creatures. Even more important, he will probably lose his ability to survive natural disasters such as blizzards, earthquakes, hurricanes, and so on. Unable to cope, this victim of severe overcrowding will die by the millions.

Calhoun predicts that man could be wiped out if this route is his choice. There will be no survivors to start over again because individual man will have completely lost his understanding of technology, as well as his ability to survive as a hunter-gatherer. Unable to use civilization's technological inventions, and unable to hunt or gather, he will pass out of existence surrounded by the riches of his ancestors' creativity. Like Calhoun's mice, man's doom could be sealed by the decision to allow the population explosion to continue unchecked.

Stagnation. The second choice is to stop population growth near the nine billion mark. This is what might happen if the Zero Population Growth (ZPG) movement should be successful. To achieve zero population growth, the number of births must be reduced to the number of deaths each year. Calhoun believes that this choice is a bad one, for the following reason: if population levels off at a point near the maximum idea carrying capacity of all mankind (see appendix), then the potential of the individual becomes constant. It can no longer grow.

But if the potential of the individual can no longer grow, human cultural evolution will have reached a dead end. Each person will play out his role in life and then be replaced at death with another who will play the same role. Man in the past has been characterized by his ability to adapt to a changing environment, and by his need to discover new avenues for creative growth. This will no longer be the case. With this choice, man locks himself into a preset pattern, never again to change.

RxEvolution. Following this choice, the population of the world is decreased, and RxEvolution results. Calhoun coined the term RxEvolution by substituting Rx, meaning prescription, for the R of revolution. The result, RxEvolution, is "the prescription for evolution." If this route is selected, man for the first time controls the design and future of his own evolution.

When the world's population is decreased below nine billion, the potential of the individual can continue to increase. Put another way, as population numbers move downward below the nine billion mark, it is possible for the maximum idea carrying capacity of mankind to expand further.

Calhoun believes that cutting the world population in half will permit a four-fold increase in the use of conceptual space. Half as many people will have to cope with up to four times as much information! Is man up to this task? Calhoun thinks so. He believes that the individual's idea carrying capacity will increase in proportion to the enlargement of conceptual space.

Thus, if man, the explorer of wider and wider horizons and the creator of newer and better ideas, is to continue his evolutionary growth, he must decide soon to reduce the world's population. If Calhoun's theory is correct, man must make his decision within the next 50 years.

RESPONSIBILITY OF THE INDIVIDUAL

Many predict radical changes for the society of the future—changes that will begin to take effect during your lifetime. For example, it was mentioned that all of the peoples of the world would make up but one single community—in Calhoun's words, a *global village.* Thus, all of mankind would be linked together by common goals and by a vast and complex communications network.

It is important to point out that international politics has nothing to do with the emergence of the global village. Something much more basic is involved. These factors are continued cultural development and man's need to survive in the face of increasingly overcrowded conditions. Thus, the logical end result of man's use of conceptual space is the single world community. We are probably closer to it than anyone realizes.

Unfortunately, a human population explosion that shows no sign of slowing down parallels the growth of conceptual space. Soon, however, this means that the idea carrying capacity of all humanity will reach a maximum. It will no longer increase but rather remain constant. At this point, man will have three choices—death of the species as a result of overpopulation, eternal stagnation because it will no longer be possible to expand conceptual space, or continued cultural evolution as a result of reducing the population.

Many experts believe that man has a two-fold need, a desire for food for the body and a desire for food for the mind. Interestingly, as far as food for the body is concerned, we remain unchanged from our primitive hunting and gathering ancestors. We still need 1500 to 3000 calories of food energy per day. Our hunger for food for the mind, however, becomes greater and greater as time passes. We must continue to create to satisfy this need.

We must also recognize that there probably is an upper limit to man's inventiveness and creativity. As we approach that point, decisions about the future must be made. These decisions belong in part to today's youth, and in greater measure to the following generations. If what we are trying to say in this book is correct, we must begin today to stop population growth, and take the necessary steps during the next 40 to 50 years to start a gradual population decline.

Can you imagine a world with no physical room for any more people, and no more conceptual space for cultural growth? We haven't reached these dead end points—yet. Whether we do or not, and what happens if we do, is the responsibility of every man, woman, and child in the world.

APPENDIX

Biologists calculate the total amount of living matter—the *biomass*—in a population by using the expression $B = Nw$. The B in this simple formula represents biomass; w is the weight of the average animal in the population, and N is the total number of animals. Thus, the total amount of living matter in a population is simply the product of the total number of animals and the average weight of an individual animal.

During his research on the future potential of man, Calhoun noted the similarity between biomass and a concept—*ideomass*—that he was developing. Ideomass is the "idea carrying capacity" of mankind referred to in Chapter Nine. In developing the ideas of senescence, stagnation, and RxEvolution, Calhoun then proposed the relationship $I = Nd$, in which N is the total population of the world, d is the potential of the average individual, and I is the idea carrying capacity of all mankind. He suggests that there is a point in human population growth at which the idea carrying capacity I reaches a maximum, and can no longer increase. He sees this point being reached when the total population is about nine billion. Thus, when I in $I = Nd$ is constant and cannot increase, only two possibilities exist. If the total population N increases, then the potential of the individual d must decrease. On the other hand, if N decreases, then the potential of the individual d can continue to increase.

A simple analogy may help you understand the meaning of the relationship $I = Nd$. Imagine that the maximum idea carrying capacity (maximum I) of all mankind is represented by the volume of a large box. Now picture the box filled with balls about the size of Ping-Pong balls. Each ball represents an individual human being, and the volume of a ball that person's potential (d). The number of balls (N) in the box represents the population of the world (about nine billion) when the point of maximum idea carrying capacity is reached.

Now, the volume of the box cannot change—it represents maximum idea carrying capacity. But suppose the population, that is, the number of balls in the box, was to increase. There is only one way this can be done—*the balls must be smaller* to get more of them into the box. But since the volume of each ball represents the potential of one individual, we see that when the population goes above nine billion, the potential of each individual decreases. When the population decreases, however, the potential of the individual increases. In terms of the volume of the box, when fewer balls are in the box, *they must be larger* to completely fill it. This larger ball volume represents a greater potential for the individual.

ACKNOWLEDGEMENTS

Photographs and illustrations in this book are reproduced courtesy of the following:

cover: Ellis Herwig, Stock-Boston
page 14: Wide World Photo
page 29: Ringling Bros.—Barnum & Bailey Combined Shows, Inc.
page 44: Dr. John B. Calhoun
pages 48-49, 59: from "Population Density and Social Pathology," John Calhoun. Copyright © 1962 by Scientific American, Inc. All rights reserved.
page 57: Robert Weber, The New Yorker Magazine
pages 84-85: Dr. Garrett C. Clough
page 93: Michelle Stone/Editorial Photocolor Archives
page 107: Authenticated News International
page 115: Yoichi R. Okamoto
pages 116, 127: from "RxEvolution, Tribalism and the Chesire Cat," John B. Calhoun, URBS DOC No. 167, January, 1971.

INDEX

Adrenal glands,
   function of, 83, 86-93
   of lemmings, 83
   of man, 90-93
   and overcrowding, 86-93
   size of under stress
      conditions, 86
   of Sika deer, 78, 79, 83
   of woodchucks, 89, 90
Adrenal hormones in man,
   and overcrowding, 92, 93
   and sensory perception,
      91-93
Aggressive behavior in
      man, and
      overpopulation, 100,
      102
Animal space
      requirements, 27, 30,
      31, 42
Antisocial behavior, in rats,
      46-47, 50
Attack distance, 28, 30

Behavior and crowding, 16,
      17, 36
   in langur monkeys, 70-74
Behavior and stress,
   in man, 20, 36
   in rats, 43, 45
Behavioral sink,
   in langur monkeys, 72-74

in man, 51
in mice, 54, 117
in rats, 50
Biomass, 133
Breder, C. M., 40
Bruce Effect,
   in house mice, 80, 81
   in meadow voles, 81, 82
   and overcrowding, 82
Bruce, H. M., 80

Calhoun, John,
   conflict experiment with
      rats, 58-61
   mouse population study,
      113-119
   rat experiments, 42, 43
   view of human evolution,
      120-125
California voles, population
      control in, 41, 42
Care of the young in man,
      and overpopulation,
      100-102
Christian, John,
   Sika deer experiment,
      76-79
   woodchuck experiment,
      89-90
Competitive behavior,
   in human females, 97-99
   in human males, 97-99

137

141

ABOUT
THE
AUTHOR

"Lee" Drummond is well known in the field of science education and has written several other science books for young readers. His numerous articles on science teaching have appeared in journals such as *Science and Children, Science Activities, School Science and Mathematics* and many more. He is also the author of several popular science articles that have appeared in *Sea Frontiers, Science Digest,* and *Yankee.* Mr. Drummond earned his bachelor's degree at Bowdoin College, and holds graduate degrees from Hofstra and Wesleyan Universities. He is editor-in-chief of the school science department for a major textbook publishing firm.

Sailing, one of Mr. Drummond's hobbies, is the subject of one of his books; other interests include restoring and building clocks. Mr. Drummond and his family live in Bedford, Massachusetts.

DRAGON CITY

DRAGON CITY

By Katie and Kevin Tsang

Fountaindale Public Library District
300 W. Briarcliff Rd.
Bolingbrook, IL 60440

STERLING CHILDREN'S BOOKS
New York

To Kevin's mom, Louisa,
and Katie's mom, Virginia.
Thank you both for everything.

STERLING CHILDREN'S BOOKS
New York

An Imprint of Sterling Publishing Co., Inc.

STERLING CHILDREN'S BOOKS and the distinctive Sterling Children's Books
logo are registered trademarks of Sterling Publishing Co., Inc.

Text © 2021 Katie Tsang and Kevin Tsang
Cover illustration © 2021 Pétur Antonsson
Interior illustrations © 2021 Vivienne To

All rights reserved. No part of this publication may be reproduced, stored in a
retrieval system, or transmitted in any form or by any means (including electronic,
mechanical, photocopying, recording, or otherwise) without prior written permission
from the publisher.

ISBN 978-1-4549-3600-8
978-1-4549-3601-5 (e-book)

Distributed in Canada by Sterling Publishing Co., Inc.
c/o Canadian Manda Group, 664 Annette Street
Toronto, Ontario M6S 2C8, Canada

For information about custom editions, special sales, and premium and corporate
purchases, please contact Sterling Special Sales at
specialsales@sterlingpublishing.com.

Manufactured in Canada

Lot #:
2 4 6 8 10 9 7 5 3 1

02/22

sterlingpublishing.com

Cover design by Tom Sanderson
Interior design by Julie Robine
Interior illusrations by Vivienne To

CHAPTER 1
THE MOON

The moon is patient. High in the sky, she waxes and wanes. Sometimes she smiles, sometimes she cries. The teardrops of the moon fall down into the sea, lost forever, no matter how many times she pulls the tide back and forth, revealing what lies hidden beneath the waves. The moon is always there, always watching, always waiting, until she is most needed. But only by those who are worthy of her power.

High in a spiraling tower, a dragon watches the sky, waiting for the very moment when the moon will appear. The dragon asks for forgiveness, but there is no answer. The moon does not smile, and she does not cry.

The dragon turns and gazes down at her shadow—one she no longer recognizes. Even though it is always there . . . even in the dark.

CHAPTER 2
DRAGON TEETH

Billy Chan was inside a dragon's mouth.

It was hot, damp, and smelled terrible. And, of course, there was the ever-present danger that the dragon might decide to crush Billy between his jaws.

He peered out from behind the dragon's sharp teeth. "Could you open up a little wider? There's something stuck back here," he called out.

The dragon, who had sleek black scales and flowing silver whiskers, as well as a surprising and impressively long silver beard, grumbled but unhinged his jaws a bit so Billy had more space to move around.

"Thank you!" Billy said, repositioning himself so he had better access to the dragon's back molars. Because Billy Chan, twelve-year-old surfer from California, was currently spending his days cleaning dragon teeth.

"*Ugh*," he muttered under his breath, as he picked out something with feathers from in between the dragon's back two teeth. "What *is* this?" Why couldn't dragons use toothbrushes?

A loud voice from outside the dragon's mouth interrupted his thoughts. "Now, how sharp do we want these today?" The voice belonged to Billy's friend Charlotte Bell, who spoke with a Southern accent because she was from Atlanta, Georgia. She was filing the dragon's front claws.

"Wait! Don't reply!" Billy said. But it was too late, the dragon's mouth was already moving, his teeth gnashing dangerously close to Billy's hand.

"As sharp as you can get them," said the dragon.

"And what about the back ones?" called another voice, this one in a lilting Irish accent. Dylan O'Donnell, from Galway, Ireland, was in charge of filing the dragon's back claws.

"COULD YOU PLEASE WAIT UNTIL I AM OUT OF HIS MOUTH TO ASK QUESTIONS?" Billy shouted, as he yanked his hand out of harm's way.

"Sorry, Billy!" Dylan called out. "We'll be quiet now."

There were faint murmurs of agreement, and then his friends fell silent. Billy wiped the sweat off his brow and waited. All he heard were the dragon's heavy breathing and the noise of dragon claws being sharpened, which sounded just like blades slicing against each other.

It *must* be safe to start the teeth cleaning again. Surely. Taking a deep breath, Billy moved farther back in the dragon's mouth, regretting that he'd ever offered to be in charge of teeth. With a small blade, he started picking at a piece of bone stuck behind a back tooth, careful not to stab the dragon in his gums. Billy had learned the hard way that dragons' gums were surprisingly sensitive.

"Your beard is looking particularly nice and shiny today," said a soft, musical voice. It was Liu Ling-Fei, who was responsible for scale shining, mane brushing, and whisker maintenance.

Ling-Fei had grown up in the mountains of China, near where Camp Dragon had been, which was where the four friends had met and now felt like a lifetime ago. Ling-Fei was always happy to offer a kind word to a dragon or a human. This was usually a quality Billy appreciated in her, but not right at this moment. As long as the dragon didn't reply—and he probably wouldn't—Billy would be fine. Most of the dragons they worked with barely acknowledged them, let alone responded to their praise.

"Thank you!" rumbled the dragon, clearly pleased with the compliment.

Billy groaned as a glob of dragon spit hit him in the face.

"*Guys!*"

"Sorry!" said Ling-Fei, before letting out a quiet giggle. "Although that wasn't a question. It was a compliment!"

Billy rolled his eyes as he listened to his three friends chatting as they continued grooming the dragon. Easy for them—they weren't inside his mouth.

But still, Billy knew how lucky they all were.

To be alive and to be together.

They had each other, and Billy clung to that whenever he found himself feeling sad or scared when he heard the roars of dragons and the screams of humans.

Everything had changed.

After Billy's dragon, Spark, had betrayed them and joined the Dragon of Death, giving her the eight pearls she needed to choose her own destiny, the world around them had disappeared.

When it had come back, it was completely different. Billy, Ling-Fei, Charlotte, and Dylan had awakened in a dark and distant future. One where the Dragon of Death ruled with a fearsome and terrible might. One where, somehow, she had been ruling for years and years already, even though it felt like only moments had passed between their lives in the past, in the Dragon Realm, and this version of the future, where there was no Dragon Realm and Human Realm, only Dragon City and the Void beyond. Both the Dragon and the Human Realms had been decimated and devoured by the Dragon of Death and the nox-wings and their never-ending quest for power, leaving Dragon City as the only habitable place for dragons and humans.

But at least Billy and his friends had been together, and they still had their memories of their lives before. And even though they had been separated from their dragons, they had heard their dragons when they had first arrived in Dragon City and found themselves in chains in an unfamiliar and terrifying cityscape. Knowing that their dragons were alive had given them hope. Because the dragons were more than just friends. Deep in Dragon Mountain, the four children had each heart-bonded with a dragon, connecting them forever. Dylan had bonded with Buttons, a healer dragon who cared deeply for humans. Ling-Fei's dragon was Xing, a dragon with the ability to seek out magic and power, and whose tough exterior hid a kind heart. The fierce warrior dragon Tank was Charlotte's heart-bonded dragon, and the two of them together could take on almost any opponent. As for Billy . . .

He didn't like thinking about his dragon, Spark, with her electrical powers and ability to see into the future. He had trusted her more than anyone, and she'd let him down. But, despite everything, part of him hoped that they were still connected through the heart bond. But when he tried to reach down their bond, there was nothing. It made him feel empty inside, like a part of him was missing.

Even though they were separated from their dragons, they weren't alone in the terrifying world of Dragon City. A tiny gold flying pig had been sucked into this future alongside them. And even though it couldn't speak, this diminutive pig could understand them, Billy knew; when they'd needed help escaping their shackles, Billy had asked the pig to find the key.

It was a big ask for a tiny pig, but the pig had previously brought him Dylan's Claddagh ring, after all, and it had led Billy and the others to where Dylan was trapped in a tree by dark magic. So surely it could find a key now to unlock their chains.

Hours had gone by, during which the four friends had watched in horror as nox-wings swooped down on unsuspecting human workers and tossed them up into the air in some sort of twisted game, laughing as they did. At one point, a dozen humans had run down the street, screaming that "Death's Shadow" was coming. Billy and the others had pressed themselves closely against the wall they were chained to as a giant dark shadow passed by overhead, leaving a trail of crackling cold air in its wake. They'd tried to ask for help from passing humans, but everyone stared blankly at them and then hurried on, not even daring to glance back.

"Humans are so frightened in this time," Ling-Fei had said softly. "Everyone is only looking out for themselves."

"We don't need anyone else," Billy said, trying to sound brave, even as fear nipped at his toes. "We've got each other. And the pig will find the key. I'm sure of it."

"I can't believe we're putting all our faith in a flying pig," Charlotte muttered.

"That little pig saved me before. It can do it again," said Dylan with forced cheer.

But as the sun had sunk lower in the sky, so, too, had Billy's heart. Fewer and fewer humans were passing by, and the dragons who flew overhead clearly had no interest in helping anyone. If anything, they seemed to be antagonizing the few humans hurrying through the streets.

Billy began to wonder if trusting the pig had been a mistake. If they'd be chained up here forever. Or if they'd even survive the night. He didn't have any other ideas, and he didn't know how much longer he could put on a brave face for his friends.

Finally, just before sundown, the tiny gold pig had come back! Billy was so excited and relieved that he let out a great shout of joy at the sight of her.

"*Shh!*" Charlotte said. "We don't want to draw any unnecessary attention to ourselves!" But she was grinning, too, and relief shone from her eyes. Ling-Fei clapped with delight as the pig came closer, and they saw that she was carrying a small key in her mouth.

"You, my friend, are a genius," said Dylan, holding out his hand as the pig dropped the key into it.

"Thank you," Billy said with a wide grin.

"I wasn't talking to you," said Dylan pointedly. "I was talking to the pig. A genius flying pig!"

"I'll never look at bacon the same way again," added Charlotte with a laugh.

"Don't say the B word in front of the pig!" whispered Ling-Fei, gently stroking the back of the tiny gold pig.

"Let's try the key," said Billy, holding his breath. With trembling fingers, he put the key in the lock, and, with a click, the chains fell off his wrists. Moments later, they were all free.

As the bells began to toll, the sky grew dark and the city's neon lights flickered on. Billy, Ling-Fei, Charlotte, and Dylan crept down darkened streets and alleys, careful to stay out of sight, and the tiny gold pig fluttered anxiously around them. Billy hadn't known it at the time, but any human caught out in the streets after dark was considered fair game for a nox-wing to take as a slave, a snack, or, worst of all . . . to drain of life force. But he'd felt a primal fear at being out after dark with the bright lights and darkening sky closing in on them like snapping teeth. In a panic, Billy found himself reaching out through his bond to Spark to see if she was in this future as well. And he'd felt something answer back.

But it hadn't sounded like Spark. It had felt like a thread pulling him in a certain direction. And with nothing else to go on, he'd followed it. "Something is telling me to go this way," he said.

"Something or someone?" challenged Charlotte, frowning. "What if it's . . ."

"It's not," Billy said brusquely. He knew what Charlotte was implying. That Spark, his Spark, could no longer be trusted. But an invisible force was urging him on, and he had no choice but to follow it. "Do you have any better ideas?"

So Billy led his friends through the unfamiliar streets, careful not to be crushed by huge dragon feet or caught up in dragon fire or frozen by dragon ice. They avoided the snapping jaws of bickering dragons and gasped at the sight of humans, so many humans, all with their heads down, doing whatever the jeering nox-wings demanded. Billy noted that not *all* the dragons seemed to be antagonizing humans. Some of them were being shouted at by nox-wings too. These dragons wore collars crackling with electricity, and Billy knew that meant they were being controlled by dark magic. But although it was terrible to see, it also gave him another slight burst of hope that there were still some good dragons in this future.

The streets were lined with huge buildings, each big enough to house dragons. They had gigantic windows and doors that dragons flew in and out of. And in the center of the city, a towering skyscraper stretched high into the clouds. It radiated with pulsing electricity, and purple smoke poured out of the windows.

"We should stay far away from that tower," said Ling-Fei, eyeing it with unease. "It feels evil."

The four friends hurried through the dark, careful to avoid any dragons, until they turned down a narrow alley and reached a sewer grate at the end. Billy paused. The feeling pulling him onward was stronger here, drawing him closer, like someone cold seeking heat.

He looked at his friends and swallowed. "I think we should go underground."

A voice from the dark slithered out. "A good idea, boy, a good idea indeed." And then a tall, thin figure emerged from the shadows. It was a woman in a black cloak with silver hair down to her waist, but her face was young and smooth. Most alarming of all was the glowing knife gripped in her hand and the net thrown over her shoulder. She caught Billy staring and flashed him a sharp grin.

"That's right, I'm a nox-hand. I could get a fine reward for turning in humans, especially *young* humans, caught out after dark." Without having to ask, Billy knew that a nox-hand was the human version of a nox-wing—someone who dabbled in dark magic and served the Dragon of Death.

Charlotte stepped forward, her hands on her hips. "I'd like to see you try. In case you can't count, there are four of us and one of you."

"Charlotte," hissed Dylan in a high-pitched whisper, "she has a *knife*. And we don't have our pearls!"

Charlotte shrugged, keeping her gaze on the cloaked woman. "I could still knock her flat on her butt."

Billy moved closer to Charlotte. If she was going to fight this nox-hand, she wasn't going to do it alone. Without saying a word, Ling-Fei did the same, and after a short moment even Dylan stepped forward, muttering under his breath. This woman didn't know what they had faced. They had battled a giant scorpion, conquered nox-wings, and even defeated the Wasteland Worm. They were not afraid of a stranger with a cloak and a knife.

At least not *that* afraid.

The nox-hand took another look at them and then laughed, long and loud. "Perhaps I won't turn you in. It would be a shame for all that energy to be sucked up into the Tower."

Billy glanced anxiously up at the glowing Tower that loomed over them. Was it functioning like the red dome had back in Dragon Realm—sucking up life force and energy to be used for dark magic? The idea chilled him to the bone.

The nox-hand kept talking. "Run along, little rodents, before another nox-hand finds you. Or worse, a nox-wing. They won't find you as amusing as I do."

She turned on her heel and strode into the dark, holding her knife aloft. When she reached the end of the alley, an enormous turquoise dragon flew around the corner. Billy watched as it went straight for the woman, but as it registered her glowing knife, it stopped short and gave her a curt nod. The woman nodded back and disappeared into the shadows. The nox-wing raised its large head and looked straight at Billy. Its tongue flicked out, as if it were licking its lips.

Billy gulped.

"Come on!" he said to his friends. He didn't want to wait to see what would happen if the nox-wing reached them. He yanked open the sewer grate. "We've got to get out of here!"

"We don't know what's down there!" cried Dylan.

"It can't be worse than what's heading straight for us," said Charlotte. "I'm going in." And with that, she slid into the dark.

There was a muffled thump. "It isn't too far a drop," she called out. "But it is wet down here. And gross."

"Hurry," said Billy to Ling-Fei and Dylan. The nox-wing was forcing its way through the narrow alley, its snapping jaws getting closer with every second. They were running out of time. "I'll go last."

"Be careful, Billy," said Ling-Fei. Then she glanced at Dylan, who was standing frozen, staring at the nox-wing as it grew closer and closer. "Come on, Dylan!" She grabbed Dylan's hand and pulled him along with her. With a loud squeak, the tiny gold pig flew in after them.

Then Billy sprang into action. He slid into the sewer,

feetfirst, pulling the grate behind him. It slotted back into place with a clank as he let go and dropped down.

A second later the nox-wing crashed snout-first into the grate. With a roar of frustration, it ripped the grate away, but it was far too large to fit through the narrow opening. It shot a jet of fire down into the sewer, momentarily lighting it up. Dylan screamed as they scrambled away from the flames, but Billy took advantage of the brief moment of light to take in their surroundings. They were in a long tunnel with rusted tracks lining the ground, and the walls were slick and damp.

"Stay underground where you belong, vermin!" the nox-wing shouted with a laugh, and then it flew away into the night.

CHAPTER 3
THE SECRET TRAIN

They wandered the tunnel for hours. Whenever they heard footsteps or saw shapes, they hid in the shadows, careful to avoid being seen by anyone. From their encounter with the nox-hand, they knew that even other humans weren't to be trusted in this future.

Billy had continued to follow the thread, pulling him farther and farther into the tunnels beneath the city. It had been dark when they'd initially jumped into the sewer, but as they walked on, Billy noticed that the walls of the tunnels had started flickering with light. Soon neon lights pulsed brightly all around them and seemed to be heading toward the center of the city, where they'd come from. But the light wasn't contained in glass or bulbs; instead it streamed directly along the walls in a thrumming current, like a flowing river of electricity.

And Billy felt as if he were inside a wire connected to a power source—electricity flashing all around him.

And then, right when Billy had felt the strongest pull, they'd hit a dead end. The tunnel stopped at a solid concrete wall. Dylan cleared his throat. Ling-Fei paused, looking around carefully, and Charlotte sighed loudly. Billy's heart sank. He'd led them all the way down here for nothing.

But then he remembered another dead end he'd encountered. One that hadn't been a dead end at all. When he'd followed his friends into the mountain behind Camp Dragon, what he thought had been a solid gray wall had actually been a door to an enormous cavern where his friends were. Where their dragons were.

Perhaps this wall wasn't what it appeared to be.

"Maybe we can go through this wall . . ." Billy said.

He wished Ling-Fei still had her pearl power to sense magic and life. He wished Charlotte still had her superhuman strength. And he wished he had his agility power. But all they had were each other, their wits, and their courage, and that would have to be enough.

"It wouldn't be the weirdest thing that has happened today," Dylan said with a wry smile. "We *did* see the sun fall out of the sky and crack like an egg. And then a turtle called the Destiny Bringer crawled out of it and made the whole world disappear. So, sure, maybe we can go through this wall."

"But we should be cautious," said Ling-Fei.

"Y'all know I hate being cautious," said Charlotte.

She reached out and slapped a hand on the solid wall. "Yep. That is a real wall. What's the next bright idea?"

"Maybe there's a trapdoor somewhere," suggested Ling-Fei. All four of them began to feel around on the wall, hoping their fingers would snag on some kind of lever or mysterious button.

Nothing.

The tiny gold pig squeaked and landed on Billy's shoulder. Even she looked exhausted; her usually bright eyes were cloudy with fatigue.

"I think that pig just yawned," said Charlotte. She followed suit, opening her mouth wide and stretching. "Maybe we sleep here? I'm exhausted and starting to get delirious."

"No way," said Billy. "It's too dangerous. We have to get somewhere safe."

"And you think that 'somewhere' is behind this very sturdy concrete wall?" asked Dylan.

"I know it is," said Billy, his voice coming out sharper than he intended. "It has to be!"

Dylan, Charlotte, and Ling-Fei exchanged a look and Billy stared up at the seemingly impenetrable concrete wall in front of him. "Why can't you all just trust me?"

"Billy, you didn't tell us that Spark was starting to go dark," said Ling-Fei gently. "You kept a huge secret from us and the dragons."

"And look what happened!" said Charlotte.

"I thought I was doing the right thing," said Billy, glaring at his friends. But he knew they were right.

"We aren't mad at you," said Ling-Fei.

"I'm a little mad," interrupted Charlotte. "We don't have our pearls or our powers or our dragons because Billy kept a secret from us! And now we don't know how we'll ever get home."

"It isn't all Billy's fault," said Dylan. "I think if we're blaming anyone, we can blame the Dragon of Death, right?"

"I'm sorry, okay? I'm sorry!" cried Billy. His voice echoed in the tunnel and Charlotte whirled around, putting her finger to her lips and shushing him.

"Okay! Okay! I accept your apology! But be quiet! We don't know what else is roaming through these tunnels."

"Which is exactly why I want to find us a safe place to rest," said Billy. "And I know I messed up before, but something is drawing me to this wall. It's telling me that we need to get through it." He looked at the train tracks they'd been walking on. They stopped at the wall, but in a way that seemed as if the wall hadn't always been there.

"What if that something is Spark?" said Dylan, eyeing him carefully. "You can't trust her anymore, Billy."

Billy sighed. "I know." He ran his fingers through his hair. As he did, he felt the prickle of static shock. But instead of stopping at his fingertips, it whizzed up his arm and then through his entire body. It felt like he was buzzing with his own electricity.

"That was weird," he muttered, momentarily distracted from his argument with his friends.

"Billy, I think we need to turn back," said Charlotte.

"I agree," said Ling-Fei. "There isn't anything here."

"Wait," said Billy, an idea slowly forming. He stared up at the racing electric current running along the walls of the tunnel and noticed that it appeared to keep going through the concrete wall in front of them.

They'd been through portals under lakes and ones that they'd created themselves using the power of their pearls. This was another kind of power. Maybe, just maybe they could use this energy as a kind of portal. . . .

"We should hold hands," he said. "Like that time we opened the mountain."

"We can try that, but if it doesn't work we're heading in another direction, okay?" said Charlotte.

Billy nodded. Charlotte took his hand, Ling-Fei grabbed onto her other one, and Dylan held Ling-Fei's. They waited a moment, all holding their breath in anticipation. Then Billy felt another flicker of static electricity run through his hand into Charlotte's. She flinched, and so did Ling-Fei and Dylan.

"What was that?" said Dylan, sounding panicked. "Is something happening?"

"It was just a shock," said Charlotte. "This isn't working."

"I think *something* might be happening," said Ling-Fei, staring down at their interlocked hands as another shock flickered through them.

Billy felt a pull toward the current of energy streaming on the wall next to them. It was as if the current were calling to him, as if the thread that had led him here wanted him to reach out to it. "I need you guys to trust me," he said. "I'm going to try something."

"WAIT! WHAT ARE YOU DOING?" yelled Dylan. "I ABSOLUTELY DO NOT TRUST YOU!"

But Billy wasn't listening, he was reaching with his free hand toward the tunnel wall and the glowing current of power flying across it. He hoped his instinct was right and that he wasn't about to do something incredibly stupid.

"Hold on!" he shouted. And then he stuck his hand straight into the current of light and power.

His whole body lifted up off the ground and was sucked into the current, along with his friends. All around him was light so bright that he had to close his eyes against it. And then he felt a prickling heat like static shocks all over his body. And then it was as if he were flying. But it wasn't like flying on Spark, it was as if he, Billy Chan, were flying through a rainbow current of light and power, with energy crackling all around him.

When he finally dared to open his eyes, Billy began to laugh, because now he could see exactly where they were. They had traveled through the power line to the opposite side of the concrete wall and below them was an empty subway station. Here, he knew, was where they needed to be.

Still holding on tight to Charlotte, he leaped out of the current as if he were leaping out of a glowing, fast-moving river, and they all crumpled to the ground. Charlotte stood first, her hair sticking up in every direction.

"HAVE YOU LOST YOUR MIND?" she shouted.

Billy was shaking, but he was grinning too. They'd done it!

"We survived, didn't we?" he said.

"You didn't know that!" cried Dylan.

"We had to try!"

"Everybody calm down!" said Ling-Fei in a surprisingly loud voice. "We have to make sure we're all okay!"

Billy realized that his own hair was also standing on edge. Dylan's glasses were sparking, and he looked dazed. "I'm okay. What just happened? And where are we?"

"Well, I'll give Billy credit for this—he got us through that concrete wall," said Charlotte. "And it looks like we're the only people in here." They gazed around the empty, old subway station. All they could see was a rusty-looking train with four carriages and, beyond that, another concrete wall.

"We traveled through the energy current," said Billy.

"But HOW?" said Dylan, staring up at it.

"I'm . . . I'm not really sure," Billy admitted. "I know that the current can go through walls, and I had a feeling I could too. It might have something to do with my connection to the Lightning Pearl."

Charlotte smoothed down her hair. "That was a pretty big risk for something you weren't very sure about."

"It felt right. The same way that jumping in the portal in the Frozen Wasteland felt right."

"I wonder if it has something to do with your heart bond to Spark," Ling-Fei mused. "Her primary power is electricity. Is it possible you've picked up some of her power?"

Billy looked down. He had a feeling that was it, but he didn't want to say it out loud. He didn't know what it meant. "Maybe." Then he looked back up at his friends.

"You guys don't think that I'm turning nox, too, do you?" Now that he'd seen a nox-hand, he knew it was possible. And even though he wouldn't admit it to his friends, it terrified him.

"No," said Charlotte adamantly. "I'm mad at you for keeping secrets from us, but you aren't evil. We'd know if you were. We would be able to feel it."

"You're still the Billy we know," said Ling-Fei with a smile. "Following your instincts and doing whatever it takes to protect everyone, even if it means taking huge risks!"

"Literally jumping into the unknown," added Dylan, shaking his head. But he was smiling too.

"But I thought I knew Spark's heart," said Billy quietly, staring back down at the ground. "I trusted my instincts about her . . . and look what happened."

"Billy," said Ling-Fei. "Spark is an extremely powerful dragon. She was probably purposefully closing off parts of herself so you wouldn't be able to tell how she was truly feeling. I know that must be painful to think about, but what happened with Spark isn't your fault."

Billy rubbed the back of his neck. "Do you guys really think that?"

"Yes," said Ling-Fei.

"Obviously, we don't blame you for *everything*," said Charlotte, rolling her eyes at him. "That would be ridiculous! But you should have told us what was going on with Spark. Friends don't keep secrets from friends. Especially gigantic, dragon-size secrets."

"We could have helped," said Dylan.

"I'm sorry," said Billy. "No more secrets. And I'll tell you guys about any ideas I have before I jump in and drag you all with me." Then his gaze landed on the old train parked a little farther down the tracks. "Right now I think we should check out that train. It looks like a place we could rest."

"That's definitely an idea I can get behind," said Dylan.

"Me too!" agreed Ling-Fei.

"I can't wait to take a nap," said Charlotte with a gigantic yawn. The tiny gold pig, who had somehow stayed on Billy's shoulder, yawned too.

As they drew closer to the train, for a moment Billy could have sworn he sensed Spark. He couldn't explain it, yet he felt that she was close. But that was impossible! Still, he tried to speak to her down their bond.

Spark?

There was no reply, but he felt a nudge to go inside the train.

The doors were sealed shut. "Hmm," said Billy.

"Surely, if we just leaped into an electric current and traveled in it through a concrete wall, we can break into a decrepit train, right?" said Dylan.

"Stand back!" said Charlotte as she walked up to the train. She quickly ripped off a strip of fabric from the bottom of her gray tunic and wrapped it around her elbow. Then she smashed her protected elbow through the train window. The glass shattered and Charlotte curtsied. "I may not have my pearl power any longer, but I'm still pretty strong."

"Well done!" cried Ling-Fei.

The pig flew in first, her tiny gold body bobbing around in the dark train. She drifted back over to the window and squeaked, beckoning them to follow her.

"That little pig is indestructible," said Dylan in awe. "I can't believe she came with us through time and then again through the electric current! I think we should make the pig our mascot."

"Wouldn't our dragons be offended?" said Charlotte with a laugh.

"I think our dragons would be *more* offended if we made them into mascots. At least Xing would," said Ling-Fei. Then she sighed. "I hope we can reunite with our dragons again."

"We will," said Billy firmly. Then he swallowed. "Or at least you all will. And I'll come along."

"Our dragons might not be your heart-bond dragon, but they're still your dragons too," said Ling-Fei.

Billy nodded, unable to speak because of the sudden lump that had formed in his throat.

"Well, let's check out this train!" said Charlotte with forced cheer. "Can y'all help me in?"

"Careful of the broken glass," said Ling-Fei as she knocked off the rest of the glass around the edges of the window with her shoe.

Once the window was clear of glass shards, Billy helped Charlotte, Ling-Fei, and Dylan through the window and then he clambered in after them. The energy current streaming along the side of the tunnel walls gave off enough light so they could see while they were inside the old train.

"Maybe we still have some of our old powers," said Billy. "I wasn't expecting Charlotte to be able to break the window like that."

"There's a lot we don't know," said Dylan. He took off his glasses and wiped them on his shirt. "I'm personally just glad my glasses have stayed enchanted to not fall off my face."

"This almost looks like a train from . . . our time," said Ling-Fei as she looked around in amazement.

"It's definitely been down here for a long while," agreed Charlotte, gingerly touching one of the seats. Twelve rows of seats lined the train car, with two on each side next to the windows and a narrow path between them.

"Should we explore the rest of the train?" said Billy, looking warily to the end of their carriage.

Charlotte yawned again, stretching her arms over her head. "Can we do that after we get some rest? I'm exhausted."

"But shouldn't we make sure it's safe?" said Dylan anxiously.

"We can sleep in shifts," said Billy. "But I think we should go in pairs, just in case. Who knows what else could get in here."

The tiny gold pig squeaked. "I know you'll be on the lookout," said Ling-Fei, holding out her hand for the flying pig to land. "But we should stay alert too."

Charlotte and Ling-Fei slept first, curled up on the old train benches, while Dylan stood at one end of the train and Billy at the other. The tiny gold pig flew between them, like a small, glowing nightlight.

It was impossible to know what time it was deep underground, so they slept on and off until hunger pangs forced them to start exploring. They'd wandered down through the carriages, staying close together, but they hadn't seen evidence of a single other person or dragon.

At the time, they didn't realize how lucky they were—to have found a safe, hidden haven. Especially when they eventually came upon a carriage with a decrepit snack bar. They'd fallen upon the bags of chips and rock-hard cookies like wolves. There were also things called "Life Bars," sealed in black plastic, and they ate those too. They offered one to the pig, and she ate the bar—black plastic wrapper and all—before they could stop her.

There was even an old toilet that somehow still flushed. They weren't sure where the waste went, so they decided to use the one farthest from their sleeping carriage. Water still ran from the taps, too, and even though they knew it wasn't safe to drink, it was nice to have something to wash their hands with. But they weren't sure how long the water supply would last, so they were careful only to wash themselves using dampened rags.

Seeing the train, with its sinks and snack bar, gave Billy a strange sense of hope. It reminded him of human life from the past. Before the Dragon of Death had risen to ultimate power.

CHAPTER 4
A DANGEROUS DISCOVERY

Discovering the secret train had happened over a month ago, but it felt like another lifetime. And now Billy spent his days cleaning dragon teeth. The first few nights the friends had hid underground, until they had no choice but to go out to find food. Without any money, or anything to barter, they'd been forced to beg shop owners. They didn't dare steal; they knew the consequences could be deadly.

One shop owner had taken pity on them, giving them food on credit. And the next time they went back, she told them that she'd heard of a dragon needing a new team of groomers. Few humans wanted that job because they were so scared of dragons, but Billy and the others knew that their experience with their own dragons would make them more comfortable, even if the dragons they were grooming were nox-wings.

They started with one dragon family and soon were grooming two or three different dragons a week. All while trying to learn more about Dragon City, listening for clues about where *their* dragons might be, and trying to stay alive in this terrifying new world they had found themselves in. They'd been so focused on surviving, on making it from one day to the next, that they hadn't been able to think about their future.

But Billy knew they couldn't go on like this. This couldn't be it. They needed a plan. They needed to find their dragons. Billy hadn't forgotten how, when they had first arrived in Dragon City, he'd spotted the flash of Xing's shimmering scales, heard Tank's mighty roar, and been soothed by Buttons's healing lullaby. Their dragons were here. Somewhere. And once they found them, they'd be able to work together to reverse what the Dragon of Death had done. They'd be able to go back home.

Home. Billy didn't let himself think too much about his family—about his parents and his brother Eddie. He didn't know if he'd ever see them again. They were still in the past . . . a past that Billy hoped he could return to one day.

But a small part of him was grateful that he didn't need to worry about his family surviving in this dark future. They may be far from him, but they were safe there.

And then there was Spark. Billy tried to keep her out of his thoughts too. But, sometimes, late at night, when he felt scared and lonely, he couldn't help but hope that Spark was somewhere nearby and that she was safe. Most

31

of all, he hoped that she hadn't gone full nox-wing, and that somewhere deep inside, she was still the Spark that Billy knew.

A sudden eruption of a truly terrible stench pulled Billy out of his thoughts and back to what he was doing in the present. He nearly gagged. "Ugh. What is that?" There was an ominous rumble from deep within the dragon whose teeth he was cleaning, and then a gigantic belch burst up from his throat, coating Billy in dragon burp stench and saliva.

"Pardon me," grumbled the dragon. "I swallowed two cows and four chickens yesterday, and I think one of them is still kicking."

Billy plucked a soggy feather out of his hair and blanched. That was his cue to leave.

"Well, your teeth are all clean at least!" he said as cheerily as he could. The dragon opened his mouth wide, and Billy jumped out.

Charlotte immediately backed away. "Ugh! You stink!"

Dylan wrinkled his nose. "She's not wrong."

"Be nice," said Ling-Fei, then her eyes widened as she got a whiff of Billy. "Oh my! That is a strong smell."

The dragon cleared his throat, and everyone stopped talking. They'd been working for this particular dragon and his family for a few weeks now, and they didn't want to lose the job. Humans who served dragons had a better chance of staying alive. The last thing they wanted to do was offend him.

Billy thought quickly. "It's an honor to be coated in your dragon burp," he said, with a bow.

The dragon raised one of his extremely furry eyebrows. The eyebrow extended off his face, and sometimes Ling-Fei would trim it. "Do not attempt to flatter or fool me, young human. You know that dragons prize honesty."

"Then I'll admit that I probably need a bath. And I'm glad you didn't decide to swallow me along with the chickens."

The dragon gave a low rumbling chuckle that sounded like a warning. "You are lucky I am not hungry right now and that I do not have much appetite for pointy-elbowed humans."

"And we don't eat humans, right, Dad?" A small dark blue dragon with shiny scales and huge horns hopped into the room. Her horns were so large that Billy was surprised the young dragon could keep her head up. "Is it my turn for grooming?" She opened her mouth wide and bared her small sharp teeth.

The large dragon laughed again, longer and louder this time. "And how do you expect a human to fit in there?"

"I'd be happy to help you out," said Billy. "Come here." He reached into the young dragon's mouth and quickly cleaned it with the grooming brush and pick.

"Your horns are looking especially magnificent!" said Ling-Fei. "Do they need sharpening?"

"Yes, please!" trilled the young dragon. "I want to look my best for the big anniversary celebration coming up."

Billy and his friends exchanged a look. A dragon celebration did not sound like a good thing for them. "What celebration?" asked Billy carefully.

"Haven't you heard? The Five Thousand Year anniversary celebration of Dragon City! Everyone is required to go. Even humans. I'm surprised you didn't receive the summons!" The young dragon gave them a puzzled frown. Billy's mind was whirling. Five Thousand Year anniversary? That meant they must be far, far in the future.

"Ohhh!" said Dylan loudly. "That celebration! Of course we know about *that* celebration. Can't wait!"

The young dragon let out a peal of laughter. "You do know that it's going to be dangerous for humans, right? The Dragon of Death and her followers will be there. Humans are there for their entertainment."

Billy gulped. He hadn't thought that they'd come face-to-face with the Dragon of Death so soon. But if *all* dragons were required to go, that meant that maybe Tank, Buttons, and Xing would be there. They couldn't miss this opportunity.

Dylan's cheeks flushed. "What I meant was . . ."

"It will be nice to see so many dragons and humans gathered together," Charlotte said smoothly. She smiled at the young blue dragon. "And we'll be careful."

"Plus, we've made lots of dragon friends as groomers," added Ling-Fei, as she finished shining the young dragon's horns.

"Dragons are not your friends," said the large dragon in a thundering voice. "Make no mistake. We keep humans around to serve and entertain us."

Something was wiggling in Billy's brain. "But . . ."

The large dragon gave him a sharp look. "But what?"

Billy had to ask. "What about the heart bond?"

"What do *you* know about the heart bond? There have not been any new heart bonds in years. Only the Dragon of Death and her consort have heart-bonded humans. It is forbidden for everyone else."

"I wish I could have a heart bond," sighed the small dragon. "I'd be so powerful! And I'd have a human to keep me company at all times."

"Even before they were forbidden, heart bonds were extremely rare," said the large dragon. He gave Billy a penetrating gaze. "I am surprised you even know about them."

"I overheard another dragon talking about it," said Billy. "When we were grooming them."

"I see," said the large dragon, but Billy could tell he didn't believe him.

"Well, we should really be on our way," said Dylan. "Lots of dragons to groom before the anniversary celebration! We'll see you there!"

"There will be hundreds of dragons there," said the young dragon with another light laugh. "You probably won't see us."

"Then we'll see you next week, as usual," said Charlotte. She nodded respectfully toward the small dragon. "I hope you have fun at the celebration."

"And I hope you stay alive!" the small dragon replied.

•

The four friends hurried out of the enormous glowing marble orb that the two dragons lived in. It didn't look like any of the other buildings or houses in Dragon City. The orb had no doors or windows, and the only way they could get in or out was to press their hands on the marble and hope it would open for them. A door would then appear the exact size of the human or dragon who needed to pass. So far, they hadn't been trapped inside the marble orb, but it made Billy nervous every time they had to go inside.

Back outside, in Dragon City, the streets crackled with electricity and every building was lit up. None as much as the Tower spiraling toward the sky in the center of the city.

Billy and his friends knew now that the electricity that charged all of Dragon City was a meld of real electricity, life force, and dragon magic. It powered everything in the city. And, most importantly, it fed the Dragon of Death's insatiable hunger for more and more power. Rumor had it that she bathed nightly in a tub of raw power, lathering herself in bolts of electricity. And the life force farms, full of humans and dragons being drained of life force, funneled energy directly to the Tower. The farms were where any human, or dragon, who defied the Dragon of Death and the nox-wings ended up.

The human workers in Dragon City tended to live by candlelight in an attempt to use clean power. Only the nox-hands used energy that they knew was probably made of life force.

Billy had been surprised at first to see that there were still shops that catered to humans, even hospitals. But then Dylan had pointed out that if the dragons wanted humans to be able to function, and to be used as workers and entertainers and not only as life force batteries, they needed food and medical care.

But the four friends only went to the shops or interacted with other humans when it was absolutely necessary. They kept to themselves as much as they could, not trusting any humans after their encounter with the nox-hand on their first night.

As soon as they returned to their secret train, going in the way they always did, through the current of power they now knew was called the "life veins," they each flopped onto a seat.

"Billy, you've got to wipe yourself down. That smell is revolting. You're stinking up the whole carriage!" Charlotte held her nose as she spoke.

"We've got bigger problems than Billy smelling like dragon burp. What are we going to do about this anniversary celebration?" Dylan tapped his foot nervously on the floor.

"Thank goodness the baby dragon mentioned it," added Ling-Fei, eyes wide. "It sounds like we have to go."

"Or we stay here and hide," said Charlotte. "Nobody knows we're here, remember? That's why we didn't get the summons in the first place."

"I KNEW IT!" cried a voice in the cabin. A voice that didn't belong to any of the four friends.

CHAPTER 5
MIDNIGHT

A chill went down Billy's spine as he looked around the train, trying to see where the voice had come from but only catching sight of his friends.

"Who's there?" said Charlotte, her eyes darting back and forth.

"IT'S ME!" the voice shouted.

Billy looked in the direction of the voice but all he saw was the pile of rags that they used to keep warm at night. "Who?"

"It's me, Midnight!" replied the voice as the young, dark-blue dragon with giant glowing horns flickered into sight.

Billy and the others stared at the dragon they had been grooming only an hour before. She looked even bigger in the train than she did in her home, where she was dwarfed by her father. Even though she was a young dragon, she was still almost as tall as Billy and twice as wide. And her horns were so tall they almost scraped the top of the train.

"How did you get in here?" said Billy, frantically looking out of the train and into the tunnel to see if there were any other dragons approaching.

"And since when do you have a name?" added Dylan.

"I *knew* you four were hiding something when you didn't know about the Five Thousand Year anniversary celebration. I could tell you were lying." Midnight shook her head. "Never lie to dragons! Lucky for you, I didn't want you to be handed in to Death's Shadow." She paused meaningfully.

While Billy had not yet seen the dragon known as Death's Shadow, he'd heard whispers of the terrifying creature who swooped down on silent wings and took humans and even dragons away for punishment, never to be seen again. "We have no interest in meeting Death's Shadow," he said to Midnight.

"That's what I thought," said Midnight smugly. "So I convinced my father that you were telling the truth but that you were just nervous, because humans are almost always nervous. But then I followed you. I can turn invisible at night—it's why other dragons call me Midnight." The young dragon grinned as her entire body disappeared so that it looked as if her head and horns were floating in midair. "Anyway," Midnight continued as her body flashed back into sight, "I almost lost you when we reached the dead end, but I grabbed onto Billy's ankle just before he traveled through the life veins. A very impressive feat—I've never seen anyone, dragon or human, do that. And now here I am!"

"A *dragon* grabbed onto your ankle, and you didn't feel it?" said Charlotte, eyes darting back and forth between Billy and Midnight.

Billy shrugged. "I felt something, but when I looked back I didn't see anything, so I thought I had imagined it."

"Why are you four hiding?" asked Midnight curiously. Then she saw the tiny gold flying pig. "And what in the Great Dragon's name is *that* thing?"

The gold pig oinked.

"Er, this is a pig," said Dylan. He scratched his head. "Do we have pigs in this time? I can't remember."

Midnight's eyes widened. "In *this* time? What do you mean in this time? And of course we have pigs. I had one for breakfast. But it was big and pink, and it definitely didn't fly!"

The gold pig squeaked in panic and flew behind Billy.

"You can't eat this pig," said Billy.

"Or any of us," added Charlotte.

"I don't eat humans," said Midnight with a sniff. "Only a starving dragon out in the Void would do that. Humans have better uses. They're worth more as life force than a small snack." Midnight's eyes widened. "Surely you must all know that. Even if you didn't know about the Five Thousand Year anniversary celebration."

"Yes, yes, we know. We're worth more in life force," said Dylan. "It's impossible to miss that one since dragons are shouting it all the time."

"Well, it's the truth," said Midnight. "Speaking of the truth, what's the real reason you didn't know about the anniversary celebration? And who are the four of you? You smell different than other humans."

Billy and the others exchanged a nervous look.

"We don't get mail down here, so we missed our summons," said Dylan. Billy started to laugh but turned it into a cough. It was true; they didn't get mail down here.

"Hmm," said Midnight, clearly unimpressed. "That doesn't explain how you can travel through life veins! Or who you are!" She looked around the train carriage suspiciously. "I've never seen four young humans without a dragon or a full-grown human looking after them."

"Our . . . full-grown human is coming back soon," Ling-Fei chimed in.

"Hmm," said Midnight. "And what do they do?"

"They're a singer," said Charlotte. "They sing for the Dragon of Death!"

"Too far, Charlotte," whispered Dylan.

"A singer!" Midnight sounded intrigued. "I love singers! I watch all of them in the Dragon Court! Which one is your human? What do they sing?"

"Son of a biscuit eater," muttered Charlotte under her breath. "Of course this dragon loves singers."

"They only sing for the Dragon of Death," said Billy, thinking quickly. "No other dragons have ever heard them. They . . . er . . . are one of the Dragon of Death's private singers. Very prestigious."

Midnight stared at Billy for a long moment, and then her horns began to glow.

"LIAR!"

"What?"

A spark shot out of the top of one of Midnight's now molten red horns. "LYING HUMANS! I CAN TELL YOU ARE LYING TO ME AND LYING MAKES ME MAD!" Another spark shot out from the other horn and blew a small hole in the ceiling.

"I'm not lying!" said Billy.

Midnight's horns were now so hot and so bright that it hurt to look at them. Billy and the others backed away.

"STOP LYING!" shouted Midnight as huge bolts of lightning blasted out of her horns and tore through the roof above them. What was once a solid ceiling was now riddled with holes.

"Okay, we might have been bending the truth a little bit!" said Dylan. "But we'll be honest with you now!"

Midnight reared her head back and roared, letting out an enormous fireball that blew the train's ceiling completely off.

The four kids ran to the far side of the carriage, trying to dodge the sparks that fell from above.

The dragon lowered her head and stared at the children. Her nostrils flared and she was breathing heavily. "I hate lying!" Midnight said between huffs. Billy watched the dragon intently as he plotted a way to escape with his friends.

But then Ling-Fei stepped forward, opened her mouth, and began to sing. She sang in Chinese, and Billy could only understand a few of the words, but the melody was beautiful. And as she sang, Midnight began to calm down, her horns slowly returning to their normal shade of silver.

After a few minutes, Ling-Fei stopped singing. "I'm sorry we weren't truthful," she said. "But it's dangerous for humans to trust dragons."

"I forgive you," said Midnight. "And I liked that song very much!"

Ling-Fei smiled. "I thought you might, as you said you love listening to the singers in the Dragon Court. It's one of my favorite songs. My nai nai used to sing it to me when I was a little girl." Ling-Fei's hand went to her neck, where she'd previously worn the jade necklace from her grandmother. The one that had, unbeknownst to her, contained the magical Jade Pearl that had given her pearl powers. Now it was in the clutches of the Dragon of Death.

"I didn't know you could sing," said Charlotte, clearly impressed.

Ling-Fei blushed. "I never had a reason to sing before."

"If I could sing like that, I'd sing all the time," said Charlotte.

"You know, I'm actually pretty musical," added Dylan. He began to whistle and snap his fingers.

Midnight frowned. "I don't like that as much. And stop trying to distract me!" Her horns began to faintly glow again.

Dylan stopped whistling immediately.

"We don't want you or your family to be in danger," said Billy.

43

Midnight let out a snort. "You're four young humans. You're not dangerous."

Charlotte bristled. "More dangerous than you think."

"Nobody knows who we are," Billy went on. "We aren't from Dragon City."

Midnight's eyes grew huge. "You're from the Void?"

Billy knew the Void was what the rest of the world outside Dragon City was called. The Dragon of Death and her nox-wings had drained all the energy in the entire world and poisoned it with dark magic, so the only place left to live was Dragon City. There was no food, no resources, no *anything* in the Void.

Sometimes humans tried to escape Dragon City, tried to survive in the Void, but they were always found by scavenger dragons. Dragons were allowed to come and go as they pleased, but very few stayed in the Void. It was too dangerous, even for dragons.

"I didn't know humans could survive in the Void," Midnight went on.

"We're from somewhere else. Not the Void," said Billy. He didn't want to lie to Midnight any more than they had to. She clearly had an instinct for when they were telling the truth or not.

Midnight frowned. "But the Void is all there is. Dragon City and the Void."

"What about the stars? And the moon?" said Ling-Fei.

"You can't be from the stars!"

"But those are other places. We're from somewhere like that. Somewhere far away," said Dylan, and Billy heard the sadness in his voice. He knew how much Dylan missed home—how much they all did.

"Is that how you were able to travel in the life veins?"

"Sort of," said Billy.

"But, Midnight, you can't tell anyone about us," said Ling-Fei urgently. "We'd be taken to the life force farms. Can we trust you to keep our secret?"

Midnight stared at each of them in turn and then grinned. "I LOVE secrets! I won't tell anyone. I promise."

As she said "I promise," the air shimmered gold for a brief moment. Billy remembered how the promise of a dragon was an unbreakable thing.

He grinned back at the young dragon. "I knew we could trust you, Midnight."

"So, now that we're sharing secrets . . . what else can you tell us about the anniversary?" asked Dylan.

As Midnight told them about the upcoming celebration, Billy felt a much-needed surge of hope. Even though Midnight wasn't one of their heart-bonded dragons, it felt good to have a dragon on their side again.

CHAPTER 6
NOODLES AND NOX-RINGS

Billy stared at the jar of eyes. The jar of eyes stared back.

"Not for humans," said the grumpy shop owner with a grunt as he picked up the jar of eyes and moved it behind the counter.

"I'm not interested in the eyes." Billy grimaced.

It was the next morning and they were buying their food for the week in between grooming jobs. Midnight had stayed at their train until late into the night, telling them about the upcoming anniversary and asking them all kinds of questions about what it was like being human. It was nice to have a dragon friend again. Even if she was a nosy, noisy dragon who couldn't control her temper.

"Although they certainly are . . . eye-catching," said Ling-Fei with a grin.

Dylan groaned. "That's terrible," he said, shaking his head.

"I don't have time for silly jokes," barked the shop owner. "Buy something or get out. It's close to curfew."

Billy placed a worn black dragon scale on the counter. "A dozen Life Bars, please."

The shop owner examined the scale and placed it under the scanner. Billy still remembered how confused he'd been when he was first paid with a dragon scale for a week of grooming work. But soon they figured out that it was the only way for humans to legally buy things in Dragon City. Humans had to use dragon scales, given to them by the dragons they served, and each one was scanned to connect them back to a particular dragon. By giving a human a dragon scale, that dragon was taking responsibility for them. Even if the only thing the human did was groom them, like Billy and his friends. It also meant that humans couldn't easily trade or barter dragon scales among themselves, because they could always be traced back to the source dragon. Not only that, but the dragons could see what humans spent their scales on. In a way, they reminded Billy of credit cards.

Gold and jewels were only for dragon use and were forbidden items for humans. Even in this future, dragons loved gold. Any human caught with a precious metal or stone would be swiftly punished and taken to the life force farms. To the south of the city were the mines, where most humans toiled. Dragons worked the mines as well, but, unlike humans, they were richly rewarded for what they found.

The mines weren't the worst place for a human to be sent, though. Worse than the mines were the factories to the east. The factories thrummed with energy and dragon magic, and they produced anything the Dragon of Death desired, as well as basic necessities for the Dragon City dwellers. Billy and the others had heard horror stories from humans who worked in the factories. Sometimes, when a machine needed more life energy, a human would be tossed in right there and then.

To the west of the city lay what was left of the natural world. This was where the animals that dragons loved to eat as delicacies were raised and where meager amounts of fruits and vegetables were harvested. Working the fields was a hard life for humans, but better than the mines or the factories.

Worst of all, were the life force farms. Directly north of the city, and under the watchful eye of the Tower, the farms buzzed and crackled at all hours of the day. Other than that, there was none of the usual noise of civilization. There, humans and dragons who had disobeyed the Dragon of Death were plugged directly into the Dragon City power supply.

Billy couldn't imagine how powerful and strong the Dragon of Death must be now. She had devoured an entire world's worth of energy, and all that was left in both realms was Dragon City. Billy knew that they would never be able to defeat her outright, and even if they did, this world wasn't fit for humans. The only way they could truly win would be to defeat her and then find a way to go back to a time before Dragon City ever existed.

Back to his time.

Back home.

But Billy didn't let himself think about his family very much. While he wasn't sure exactly how far in the future they were, he knew from the upcoming anniversary celebration and how people talked about the past that he and his friends were at least five thousand years ahead. And he knew that his family weren't here in this new timeline. They would have died a long time ago.

But Billy was determined that he and his friends would fix everything.

It felt almost impossible, but they had to. Billy couldn't give up hope.

Here, in Dragon City, most humans survived on Life Bars. Small dense rectangular bars that tasted like a mix of chalk and sludge but with enough nutrients in them for a human to live on. Billy didn't know what was in them—nobody did—but all the humans in Dragon City ate them. They were the same ones that Billy and his friends had found on the train on their first night in Dragon City.

The dragons themselves could drink from the energy fountains and only ate food as a treat. There were still some treats for humans to eat too. A few shop owners sold food Billy recognized. He'd been delighted to see that noodles had made it this far into the future. Dumplings too. He wondered why the Dragon of Death allowed such a thing, but was glad she did. Or maybe she didn't know everything that happened in her city.

"Can we get noodles today?" Dylan asked as they walked out of the shop where they had bought their Life Bars.

"Noodles are expensive," said Charlotte. "One bowl is a whole day's worth of grooming."

"What else are we going to spend our scales on?" Dylan retorted.

"He's got a good point," said Billy.

"The dragons will know what we've been buying," said Charlotte. "I don't want them to think they're over-paying us."

"We can use the Thunder Clan scale for the noodles," suggested Ling-Fei. "Midnight won't want us to get in trouble."

"And surely the dragons must have better things to do than check what food we're buying," said Dylan.

Billy's stomach rumbled at the mere thought of noodles. He grinned at his friends. "All right. Let's get some noodles."

They hurried along through the narrow back streets used by humans, until they came upon a noodle stall run by an old woman. She had an enormous gray bun, piled precariously on top of her head, and her face was as wrinkled as a walnut. Her dark eyes twinkled as the friends approached.

"I haven't seen you four before," she said with a kind smile.

They heard this a lot. The human population in Dragon City was slowly dwindling, and there were almost no new arrivals from the Void. The dragons didn't pay attention to human children, but the street vendors and other children noticed them. Luckily, all it took for sympathetic nods and clucks of concern was a story about how they'd grown up working in the factories. Mostly, though, they tried to avoid interacting with other humans. There would be too many questions that they couldn't answer.

But Billy knew this old noodle vendor was being friendly, not nosy. He smiled back at her. "We heard you have the best noodles in all of Dragon City!" He hadn't heard anything of the sort, but he wanted to distract her from wondering why she hadn't seen them before.

She beamed. "It's an old family recipe, passed down for many, many years." Her smile faltered for a moment. "Of course, some of the ingredients have changed. We don't have what my ancestors had to cook with. But I've done the best that I can."

"We'll have four bowls of noodle soup, please," said Billy, holding out the dragon scale. The old woman waved it away.

"The first bowl is on the house," she said with a wink and busied herself behind the cart, stirring a giant pot on a small burner. Billy watched as she picked up a lump of dough and began to stretch and loop it round and round again, until what had started as one long rope of dough was rows of noodles. Moving quickly, she sliced off the end and dropped the noodles into the pot. She repeated the process several times, humming to herself as she worked. Then she ladled out the soup and noodles into bowls.

She handed over the four bowls, and Billy lifted one to his mouth to take a big slurp of soup and noodles.

It was the best thing he'd tasted since they had arrived in Dragon City. The comforting taste of the broth and the chewiness of the noodles was delicious, and it reminded him of the noodle soup his dad made back at home. He felt a pang of sadness thinking about his dad, and he quickly pushed it away. He didn't have time to be homesick.

"Food!" Dylan marveled. "Real food!"

"We'll definitely be back," said Charlotte between big slurps.

"Thank you for sharing your delicious soup with us," said Ling-Fei. "You're very kind."

"Dragon City could use some more kindness, if you ask me," said the old woman, shaking her head. "So many people have joined the nox-hands, turning each other in for a little bit of glory." She clucked her tongue.

As the old woman spoke, Billy became vaguely aware of a commotion up the street, but ignored it. He knew now that nox-wings liked to start trouble. Officially, they weren't supposed to bother humans while it was daylight, unless they broke the rules.

And nox-wings were extremely good at finding humans who were breaking the rules.

But then the sound of shouting grew louder, and a burst of flame shot down the street, narrowly missing the noodle cart.

The woman whipped her head up, panic in her eyes. "Get back!" she said to the children, beckoning them out of harm's way.

"We can't let an old grandma protect us," Dylan protested to the others. "We should be the ones protecting her!"

"I heard that," said the old woman, "and I'm telling you to get back."

Billy watched as a redheaded boy, who appeared only a few years older than him, came running down the alley. No, he wasn't running, he was *flying*.

On each arm he wore two glowing bands. He had his arms outstretched, like he was leaping, and the glowing bands were lifting him up, pulling him into the air.

"Nox-rings," said the old woman. "Only given to the highest-ranking nox-hands."

Billy knew from his short time in Dragon City that dragons believed the higher one's ranking, the higher in the sky they were allowed to go. No dragon flew higher than the Dragon of Death, and no building stretched farther into the sky than the Tower. Humans, being the lowest of the low, were forced underground.

The nox-rings were made from dragon bones and power, and they granted their wearer the ability of flight. High-ranking nox-hands who had proven their loyalty to the Dragon of Death and the nox-wings were gifted them. Billy had never seen them in person, he'd only heard whispers about them.

Until now.

"If he's a nox-hand, how come there are two dragons chasing after him?" said Charlotte, pointing.

It was true. Two dragons raced through the air, quickly gaining on the boy wearing the nox-rings. It was clear he didn't know how to control his flight as he was wildly careening in the air.

"THIEF!" screamed one of the nox-wings. "DIRTY VERMIN THIEF!"

"That answers your question," said Dylan.

"Only a fool would steal a nox-ring," said the old woman with a tsk.

"A fool who is about to run into us," said Billy. "Watch out!"

In his panic, the redheaded boy looked over his shoulder at the pursuing dragons, and as he did, he lost his balance and slammed straight into the noodle cart.

The cart flipped over, noodles and hot soup spilled everywhere, and the boy lay gasping on the street.

As soon as the glowing nox-rings touched the ground, they broke into pieces and fell off his arms. The pieces lost their glow immediately and looked like an ordinary pile of broken bones.

"No!" the boy said, frantically grabbing at the pieces. He looked up at Billy and the others with pleading eyes. "Help me, please! I'm begging you! I didn't think anyone would notice that they were gone. But now the nox-wings are coming for me!"

"Death's Shadow won't be far behind," muttered the old woman, eyeing the sky. "May the Great Dragon spare us all."

A moment later, the two pursuing dragons landed with a thump, encircling the noodle cart. A crimson red dragon covered in spikes spoke first. "Foul vermin! How dare you steal what doesn't belong to you!"

Billy grabbed Ling-Fei, Charlotte, and Dylan and pushed them behind the toppled cart. Then, right as the crimson dragon roared and blew a burst of flames, he pulled the old lady behind the cart too. The cart shielded his friends from the flames, but the thief wasn't so lucky. He caught on fire and screamed in pain.

Billy couldn't stand to watch it. "Stop!" he shouted and lunged forward. But before he could go any farther, he was pulled back by the wrist. "Stay out of it, Billy," said Charlotte, who was still holding his wrist. "We can't help him."

The mustard-colored nox-wing smirked at them. "That is a wise friend you have there, groundling. You stay out of this, or we'll roast you just the same." The dragon rolled its head back toward the burning boy and, with a snort, let out a jet of water from its nostrils that extinguished the flames. "As much fun as it is watching you vermin burn, you're more valuable to us alive."

The boy was covered in horrible burns and gasping for air, but he was breathing.

"Now give us back what you stole," the mustard-colored dragon sneered.

Trembling, the boy opened his fists and the broken pieces of nox-ring fell to the ground.

"They're useless to us now!" roared the crimson dragon. "You let them the touch the ground."

The mustard-colored dragon glared at the boy. "You stole what did not belong to you and you ruined it! You must pay for your crime."

"I'm sorry!" the thief cried. "I'll never steal again, I swear!"

"The promise of a human is useless to us," sneered the crimson dragon.

Billy heard shouts from further down the street.

"Look out! Death's Shadow!"

"Run! Death's Shadow is coming!"

"Here comes Death's Shadow," said the crimson dragon to the mustard-colored one. Billy could have sworn he saw a flicker of fear in the nox-wing's eyes.

The sky darkened and Billy felt a cold wind cut through him. Then he heard a loud crack as electric bolts shot from the sky, encircling the thief and trapping him in place. Billy looked up and saw the silhouette of an enormous dragon. One with a long neck and huge wings.

A silhouette he would recognize anywhere.

No, no, no. It couldn't be.

The dragon known as Death's Shadow swooped down and landed right next to the boy. The other dragons backed away and lowered their heads to the ground.

Death's Shadow was a shimmering blue and electricity buzzed all around her. She had gossamer-thin wings and a set of beautiful gold antlers on top of her head.

It was Spark.

CHAPTER 7
DEATH'S SHADOW

Electricity flickered around Death's Shadow as she stared at the thief with liquid black eyes.

Despite his burns, the thief moved quickly, throwing himself into a low, obsequious bow, his nose pressed against the ground.

"Forgive me, Death's Shadow!" he cried. The old woman from the noodle cart slowly backed away, trembling from her head to her toes.

Billy stood rooted to the spot. He couldn't move. He couldn't look away. That dragon—the one everyone called "Death's Shadow"—that dragon was Spark. His heart-bonded dragon. He felt like he'd been punched in the stomach. *Spark* was Death's Shadow?

He wanted to cry out to her, to make her look at him, recognize him, to bring her back to the Spark he knew she was. But before he could say a word, she shot out an electric net, scooping up the thief.

The thief howled as the edges of the net brushed against his burned skin. Then Spark's own shadow moved of its own accord, separately from Spark, and Billy suddenly knew why she was called Death's Shadow. It was like Spark had split into two dragons. The shadow loomed over the boy in the net and dipped its long neck down. Billy stared in horror as he saw a flickering of the thief's life force jump from his body directly to the shadow. The thief screamed, writhing in pain, and the small flicker of light bobbed as the dragon shadow swallowed it. Billy's eyes darted to Spark, and he saw a small light burn within her, in the exact same place it had been in the shadow. And then the shadow realigned with Spark, following her movements.

Billy recoiled in horror. "How could you?" he cried out, unable to stop himself.

Spark whipped her head around at the sound and stared at Billy, as if she had never seen him before. Her eyes were so black and glassy he could see his own reflection in them. He looked small and scared.

"You dare speak to Death's Shadow?" roared the mustard-colored dragon. "You should be taken in her net along with the thief!"

Billy stood still, staring into Spark's eyes. *Please, Spark*, he thought down their bond. *It's me, Billy. You know me. And I know you're still in there.*

He waited for something, anything, even the smallest hint that she had heard him. But there was no reply. Billy wasn't even sure the bond was still there. It had been so long since he'd felt Spark. But he couldn't help himself. He had to try. And then, without a word, Spark swooped up into the sky, taking the thief with her. As she did, Billy could have sworn he felt his heart crack in two.

The mustard-colored nox-wing gave Billy a hard look. "You're lucky that Death's Shadow was in such a forgiving mood. I've seen her take humans to the Tower for less."

"My friend forgets himself," said Charlotte, yanking Billy back. "He recently hit his head and isn't thinking straight. We only desire greatness for the Dragon of Death and Death's Shadow. For all dragons!"

"A pretty speech," said the crimson nox-wing. "More humans should show such deference."

The mustard-colored nox-wing nodded.

"Well, as entertaining as this has all been, we've got jobs to do," said Dylan with forced cheer. "Dragons to groom. Very important work."

"It is indeed," said the mustard-colored dragon, twirling one of its whiskers around a claw. "Especially with the anniversary celebration coming up. Her Greatness wants every dragon looking their best."

"You're already such a handsome dragon," chirped Ling-Fei. "Surely you don't require much grooming?"

"You'd be surprised," said the crimson nox-wing with a loud guffaw. "But we've wasted enough time conversing with vermin. Be gone before our good mood sours and we decide to take you to the Tower ourselves."

"I won't forget your face," said the mustard-colored one, giving Billy a piercing glare. And then there was a furious flapping of wings, a blast of heat, and the two nox-wings were gone.

The old lady crept out from behind her overturned noodle cart. Her kind eyes had hardened. "Don't come back here," she said, struggling to turn the cart back upright. "I don't want any trouble with dragons. Or with humans who cause trouble. All I want is to sell my noodles and live in peace."

"We're sorry, lao nao nai," said Ling-Fei, using the respectful Chinese expression for the elderly.

The old woman's eyes widened. "Who taught you the old language? The dragons have forbidden it."

"My . . . my grandmother," Ling-Fei stammered. "Before she died."

"Death is a kindness here," muttered the old woman, picking up her various bowls. "But I don't care if you speak the old language. Bad luck clearly follows you. My mistake for giving you free soup."

"We can pay," said Billy, holding out the black scale.

"Repay me by never coming back here again," said the old woman. "I'm sorry. I wish I could be kind. But I have to stay alive. And I have to stay out of the farms. You understand, don't you?"

Billy nodded. "We won't bother you again. Thank you for the soup."

The old woman nodded curtly and continued to clean up the mess.

As Billy and the others walked away, he paused, looking at the broken nox-ring fragments on the ground. Without thinking twice, he quickly put a few of the pieces in his pocket and then hurried after his friends.

Back in their train car, the mood was tense. Charlotte paced up and down in the carriage. "Well, this is *terrible*. I can't believe Death's Shadow is Spark. Now that she knows we're here, she'll definitely tell the Dragon of Death. We know she's loyal to her."

"No, she won't," said Billy obstinately, kicking the old chair in front of him. The tiny gold pig flew around his head, trying to soothe him, but there was nothing the pig could do to help.

"She isn't Spark any more, Billy. She's *Death's Shadow*. Didn't you see her take some of that boy's life force?" Dylan shuddered. "She's noxious now. There's no turning back from that."

"I'm sorry, Billy," said Ling-Fei softly. "It was painful for all of us to witness, and I know it must be so much worse for you."

Billy closed his eyes tight to keep the tears from falling. He sniffed and rubbed his nose on the back of his hand. When he opened his eyes again, Charlotte was still glaring at him.

"She saw you, Billy. And that means that the Dragon of Death will know soon that we're here."

"But we aren't a threat to the Dragon of Death," mused Dylan. "Not without our pearls or our dragons."

"But I still think we're in greater danger now," said Charlotte. "What if she wants to finish us off for good?"

"Surely she has better things to do," said Dylan, waving in the direction of the surface. "Like keeping this whole city going. Devour the entire world. Plan an elaborate anniversary celebration all in her honor. That kind of thing."

Billy jerked his head up. "That's it! The anniversary celebration! We have to go."

"We didn't receive a summons, so nobody will know if we stay here where we're safe," said Ling-Fei.

"I agree with Ling-Fei," said Charlotte. "From now on, we lie low. We can groom the Thunder Clan and buy food and that is it. And you," she pointed at Billy, "you have to stay here. You're a liability."

Guilt thrummed through Billy. He knew he shouldn't have cried out to Spark, but he couldn't help himself.

"Be nice," chided Ling-Fei. "Imagine if that was Tank."

Charlotte's face softened. "I really am sorry, Billy. I know it's awful. But all the same, we can't have you causing such a commotion every time we go out!"

Something occurred to Billy. "You can't get in and out of here without me." Billy nodded his head toward the glow of the life veins streaming through the tunnels. "You need me and my connection to electricity, my connection to Spark, to be able to travel."

"Son of a biscuit eater," muttered Charlotte. "You're right."

"We stay together," Billy said firmly. "We've always been stronger together."

"I'm with Billy on this," said Dylan. He shuddered. "It was terrible when I was away from you guys."

"We need a plan," said Billy, standing up and beginning to walk back and forth in the train car. "I know I made a mistake—"

"Many mistakes," Charlotte interrupted.

"Okay, okay, many mistakes. And I shouldn't have shouted out to Spark, but I had to try something. She was taking life force from a human." Billy's voice broke. "That isn't her. It isn't." He stopped suddenly, his head drooping.

His three friends crowded close. Dylan put a hand on his shoulder, Ling-Fei rubbed his back, and Charlotte crouched down so he had to look at her.

"Billy, Spark is gone. The sooner you accept that, the better for all of us. We know it hurts, but we have to move past it."

"But you know what they say. That a heart bond can't be broken unless . . ." Billy couldn't even say it.

"Only death can break a heart bond," said Dylan in an uncharacteristically solemn voice.

"I don't want Spark to die. Even if she is noxious now, I still don't want her to die."

The four friends stood silently for a moment. Then Billy took a deep breath. "We need to figure out our next move."

"That sounds like the Billy I know," said Dylan, giving him a friendly slap on the back.

"I know we need to stay safe, but we can't stay underground like rats forever, only scurrying out to find food. We need to find our dragons; I mean your dragons . . ."

"Our dragons are your dragons," said Ling-Fei, squeezing his shoulder.

"We need to find the dragons and we need to fix this mess. Once we're reunited with them we'll be stronger. They'll know what to do."

"I know Tank is still alive," said Charlotte. "I can feel it. And he's somewhere in the city."

"I feel the same way with Xing."

"Me too," said Dylan.

Billy ignored the twinge of jealousy that his friends' dragons had been strong enough to fight the noxious pull of dark magic, unlike Spark. They were still the dragons they had always known. But he couldn't let that distract him.

"So how are we going to find our dragons?" said Ling-Fei.

Billy grinned. "We're going to the anniversary celebration."

"Even if Spark didn't recognize you in her noxious state, if those other nox-wings spot us, they'll definitely remember us," said Charlotte.

Billy's grin grew. "That's why we're going in disguise."

"I feel ridiculous," Dylan announced, staring at his warped reflection in one of the train windows.

"You look excellent," said Ling-Fei, smiling fondly at him. "But it is strange being taller than you!"

Ling-Fei sat atop Charlotte's shoulders, and a long cloak covered them both. Charlotte had lined Ling-Fei's eyes in black, put dark red lipstick on her, and pulled her hair into a tidy bun on top of her head.

Despite her reservations about going to the anniversary celebration, Charlotte had fully embraced the idea of disguises. "I love a costume change!" she declared. "And I'm glad to see my makeup skills are being put to good use even in a dragon dystopia. You look perfect—much older. And when you're on my shoulders, we'll be three instead of four. The dragons will be looking for four kids, not three adults."

Billy had been surprised when Charlotte had volunteered to carry Ling-Fei on her shoulders. Usually, she loved to be in the spotlight, but underneath the long cloak as the bottom half of the elegant lady they were hoping to pass Ling-Fei off as, she would have to be silent. And very careful not to trip.

"I'm still the strongest of all of us," she'd said.

"I don't know if I look like an actual adult," said Dylan now, still frowning at his reflection. Charlotte, as their self-appointed costume designer and makeup artist, had found him a large hat and then had had the idea to cut off some of her hair and glue it onto Dylan's face as a beard.

As for Billy, at his friend's urging, he'd cut his own hair short. He'd been proud of his hair, not that he'd ever admit it out loud, but he knew they were right. He needed a dramatic change. And with Charlotte's help he'd fashioned a moustache that tickled his upper lip.

"Now the dragons will never recognize us!" said Charlotte triumphantly.

"We'd better hope not," said Dylan, pulling his hat farther down over his head. "Otherwise, we're doomed."

CHAPTER 8
A CELEBRATION

Drumbeats thundered all around as Billy and his friends were herded toward the Dragon Court like sheep. Dragons roared from platforms that floated in midair like magic carpets. And as Billy listened more closely, he realized that the dragons perched above were actually chanting an ancient incantation—guttural snorts and bellows that flowed in sync with the drumbeats.

Every now and then, streams of fire rained down to keep the humans moving in the right direction. Billy could feel the heat from the flames as the four of them entered an enormous stadium at the center of Dragon City. The stadium, or the Dragon Court as it was known, fanned out from the base of the Tower. Above them, the Tower shot up impossibly high and disappeared into the clouds, electricity humming along its walls.

Billy had never been so close to the Tower before, and his blood ran cold, despite the flames shooting around him, as he realized that the Tower's walls were made of bones. Thousands and thousands of bones. Bones that had once been white but had been burned and charred until they were the deepest of blacks. Billy could make out human and dragon skulls peering out, the vacant eye sockets cast outward toward the city. And what Billy had thought were spikes were actually the ribs of dragons jutting out of the Tower like hooks. He couldn't bear to think where all those bones had come from. Is this where bones ended up after being sent to the farm? How could Spark support this devastation? Billy felt as if he might be sick.

"What's happening out there?" asked Charlotte, snapping Billy back into focus. "I can hardly see anything!"

"There's dark magic everywhere," said Ling-Fei, her eyes scanning the sky above them. "It's fueling everything. The fireworks, the lights—it's all powered by dark magic."

"There are more dragons than I've ever seen in one place," added Billy. "Look! Midnight and her dad are over there." He pointed to the Tower where Midnight sat perched on her father's wide back. Midnight looked excited and anxious, and her father appeared solemn and watchful. Billy almost waved at them, but then remembered they were in disguise, so he turned his attention back to the rest of the crowd. "And so many humans too! It looks like some of them are here as entertainers.

There's a whole troop of them singing and dancing in the corner of the Court."

"I have a really bad feeling that the Dragon of Death is going to expect more than singing and dancing," muttered Dylan.

Billy had been to plenty of celebrations in his life. His grandma's big 80th birthday party, the annual Chinese New Year celebrations with his family, but he had never seen anything like this.

Orbs of black and purple light floated out of the ground and into the sky. There were millions of them. The orbs of light danced in the air, as if they were performing a beautiful ballet, twirling round and round, floating higher and higher, until . . . *pop*. The orbs of light exploded with a big bang into dozens of smaller balls, like an upside-down fireworks show. Billy realized that the drumbeats that thundered all around them were the orbs of light exploding. Up, up, up the light went, bursting with a pop into more and more brilliant flecks of light that went higher and higher until everything was eaten up by the sky above. As magnificent and beautiful as it looked, it filled Billy with a sense of dread. There was something sinister about the spectacle. Even though he didn't have Ling Fei's special ability to sense dark magic, being so close to the Tower made him feel cold all over.

And then the cold feeling spread into his very heart as a cloud of darkness fell over the Court. The Dragon of Death had arrived. She landed gracefully on the raised podium in the center of the Dragon Court and stretched her wings wide. Moments later, Old Gold flew down from the top of the Tower on his own magic cloud. Billy couldn't help but scowl at the old man he had once thought was his mentor and friend. Old Gold had set up Camp Dragon with the sole purpose of finding children who could unlock the mountain, leading him to the dragons. He had been looking for one dragon and one dragon only, the Dragon of Death, and he had found her. Once they had heart-bonded they'd both gained unimaginable power.

Da Huo, the enormous orange dragon who had once been friends with their dragons but then had switched sides to serve the Dragon of Death long ago, took his place to one side of her. He had come through into this time along with the Dragon of Death and was one of the few dragons who would remember the previous time. JJ, Old Gold's grandson, had been with Billy and the others back in their own time. JJ had betrayed them, just as his dragon Da Huo had. Now JJ rode on Da Huo's back, wearing black sunglasses and an orange cape that fluttered behind him.

"I see someone updated their look," said Dylan with a raised eyebrow.

And then the crowd grew very quiet as Spark flew down from the Tower, her shadow stretching out behind her, jaws snapping separately from Spark herself. Spark's own eyes were vacant as she landed on the other side of the Dragon of Death.

The most powerful and feared dragons of all time stood before the crowd. It broke Billy's heart to see his own dragon right beside the Dragon of Death, her betrayal cutting again and again like a wound that wouldn't heal.

The Dragon of Death gazed out at her subjects, power crackling all over her body. Her sharp teeth glinted in the light as she opened her mouth to speak. The sound of her voice slithered into Billy's ears like a slippery snake, burrowing into his mind. He shuddered and tried to block out her voice, but it was impossible.

"My loyal subjects. Welcome to the most important and consequential day in our long and storied history. Today we celebrate five thousand years since I united the Dragon and Human Realms and formed the land we call Dragon City. It has been five thousand years since the Great Sacrifice was made and I began building the Tower that stands before you. Today is a day for celebration. We will, of course, indulge in the usual entertainment for such occasions—human sacrifices, life force feasts, and rare gifts to grace your hordes. But today will be special. It is not just a momentous day for dragons, but it is also a momentous day for our weaker counterparts—the humans. Today, dragons and humans alike will celebrate, because I have created something that will bring us closer together. I have created something that will make us more powerful than ever. I have created a new bond—a better bond that is even more powerful than the heart bond of olden times. It is a bond that grants powers greater than anything you have ever seen or heard of before. And it is a bond that can be made between any human and dragon who are worthy."

Roars of excitement came from above and dragons shot bursts of fire up into the air in celebration. Billy glanced at his friends in confusion. That wasn't how the heart bond worked.

The Dragon of Death's mouth split into a wicked grin, and she licked her lips. "This new, more powerful bond is not based on the heart, which is vulnerable. It is based on our true strengths—intelligence, power, and agility. The stronger you are in each of these areas, the stronger the bond, and the more powerful the result. With this new bond, we will create the most powerful dragons and humans to have ever walked the realms. But only the strongest and most deserving will form this new bond, and I will hold a grand tournament that begins tomorrow to discover the humans who have what it takes. This tournament will test those who are capable in the key areas that power the bond."

The humans in the arena cheered. A heart-bonded human would be safe and protected in Dragon City. The Dragon of Death was offering a rare lifeline, a way not only to survive, but to thrive. Billy's mind was whirling. How could the Dragon of Death have created a new bond? And why would she?

But then she said the worst thing of all. "Of course I will only allow my most loyal and trusted dragons to bond with the humans who survive the tournament. And for the human who proves themselves the best of all, well, they will be rewarded with something any human would desire. They will be bonded with none other than Death's Shadow."

Spark stepped forward, head down and eyes blank, her own dark shadow stretching out even larger than her body.

Billy felt all the air rush out of him in a whoosh. He felt dizzy, as if he might fall over.

Spark was going to be bonded with another human? How could that be? Even though Spark was noxious, even though she'd betrayed them, she was still *his* dragon. She couldn't bond with another human. Could she?

And yet, here was the Dragon of Death declaring that it would be done. In Dragon City, the Dragon of Death's word was law.

The Dragon of Death continued speaking, but Billy had stopped registering her words.

"Now, to celebrate my glorious announcement, it is time for one of my favorite games. The Dragon Death Drop starts now!"

Billy was vaguely aware of loud shouts going on around him, but it all sounded like it was coming from far away. He felt as if he were underwater and nowhere near the surface. The voices around him were distorted and the faces were blurred.

Someone shoved him and he stumbled, hitting the ground hard.

That snapped him out of his thoughts.

Then he heard Charlotte screaming. "Billy, watch out!"

He looked up and straight into the red eyes of a metallic copper–colored nox-wing, who with a snarl scooped him up with his claws, cackling as it raced high into the sky.

CHAPTER 9
THE DRAGON DEATH DROP

The copper-colored dragon flew higher and higher into the sky, Billy gripped in its front claw. He tried to kick and wriggle his way out, but it was no use—the dragon's grip was as firm as stone. And even if he did wriggle free, they were so high in the sky now that he'd surely fall to his death. Fear made his palms sweat and his heart pound, but he forced himself to stay calm, to fight the fear before it took over.

Up and up they went. Billy had never been so high in the sky in Dragon City before, and despite his fear, he couldn't help but take in the city around him. The buildings surrounding the Tower were all low to the ground— shops that were open for humans during the day and houses for high-ranking nox-hands who weren't required to live underground. But farther out from the center, Billy could see the strange structures where many dragons lived up in the clouds. At first glance, they looked like giant floating buildings, but underneath each one was a thin stream of electricity that held the buildings up in the sky. There were hundreds and hundreds of them. From afar, they looked like a field of bizarre, misshapen flowers, a wild and strange bouquet of all shapes and sizes.

And still they went higher and higher, even higher than the Tower.

"Spread out your arms and legs when you are dropped," said the copper-colored dragon. "You will fall slower, and you might actually live."

Before Billy could ask what the dragon meant, he was flung from the dragon's claws. He felt a sharp sting. He looked down and was surprised to find that he was strung upside down by his ankle on what looked like an electric clothesline. Billy saw dozens of other humans either already hanging upside down or being thrown on the line. A pair of glasses fell off the face of the person next to him. Billy watched as they fell until they seemed to disappear. They were so far up, Billy lost sight of them before they reached the ground.

Blood rushed to Billy's head as he bounced up and down on the electric line. It felt as if he were being held up by his ankle with barbed wire and he thought he might be sick. He managed to take a few deep breaths and blinked tears from his eyes as he tried to get his bearings. Farther above them, more than a hundred dragons hovered like a swarm of bees.

Billy felt a rush of wind as he saw the Dragon of Death flap into view. "I believe you all remember the rules of the Dragon Death Drop," she hissed. "When the humans drop, you must wait until my command before you can dive to retrieve them. Anything you catch in the air you can keep!"

The Dragon of Death let out a cackling roar. "Let the Dragon Death Drop begin!"

The electric line wrapped round Billy's ankle disappeared and he fell out of the sky. Wind rushed so fast and hard around him that Billy could hardly hear himself think. He forced himself to open his eyes. Humans were falling all around him. He was so high up, he could barely make out shapes in the city below him. There must be something he could do. He had to find a way to slow his fall. He clenched the sides of his tunic and stretched his arms and legs outwards like a flying squirrel. The wind caught the fabric and Billy felt a jolt as if he'd opened a parachute. It was working! He fought hard to keep hold of the fabric in his hands. He could see that other humans were falling faster than him, closer to the ground below. Billy heard a *bang* from above and he looked up to see that the dragons were no longer hovering where he had been strung up by his ankle moments before. They were diving, claws outstretched and teeth bared.

A gust of wind knocked Billy sideways, and he lost his grip on the tunic. He began to fall again like a doll in the sky, but an idea quickly struck him. Maybe he could steer himself away from the pursuing dragons. Maybe he could get away. He turned himself back into position and, to his surprise, he found he was able to steer himself in the sky. No, it was more than that, *something* else was steering him. Keeping him in the air. How could that be? Before Billy could question it any further, his body veered away from the Tower, away from the diving dragons. He was still falling toward the ground, but maybe he could find a soft enough landing to survive the fall.

Billy heard a roar from above. He looked up and saw that the copper-colored dragon had also veered off course, diving directly toward him. With a great swoosh, the copper-colored dragon scooped Billy into its claws. He was captured again.

"You are a lucky human," the copper-colored dragon snarled. "Lucky I am so fast."

But this nox-wing hadn't been *that* fast. Billy had seen it diving for him, too far to reach him, and it was the mysterious gust of wind that had kept him afloat. Not that Billy was about to tell the nox-wing that.

Billy looked down at the other humans who were in the Death Drop. In horror, he realized that most weren't going to make it. A few of the humans had managed to slow their falls enough so that dragons diving for them might just catch them if they were lucky. He closed his eyes to save himself from the gruesome scene, but it didn't save him from hearing the dull *thud, thud, thud* of the fallen. Bile rose in Billy's throat, and he gagged, vomiting all over his feet and into the air.

"You are weak. Weak and disgusting," muttered the copper-colored dragon, tossing Billy to the ground as they drew closer and then landing next to him with a thump, sending up a gust of dust in its wake. "But lucky all the same."

They were back in the Dragon Court. All around them dragons cheered and the humans who hadn't been scooped up for the Dragon Death Drop cowered and bowed.

"My Death Drop Dragons! If you caught your human, they are yours to keep," cried the Dragon of Death. "Do with them what you like—as a pet, a servant, or a source of life force."

To keep? Billy couldn't let this copper-colored nox-wing take him home! He thought falling to his death would have been the worst thing that could have happened to him, but maybe that would have been a better fate than what would happen if he were taken as a nox-wing pet.

He had to get away.

The copper-colored nox-wing laughed. "If I had known we could keep our prize, I would have chosen a human with more meat on their bones," the dragon said. "But you will do."

"I . . . I . . . need to go to work," Billy stammered. "Other dragons will miss me."

The nox-wing put his face very close to Billy's and Billy held his breath as the dragon studied him and then inhaled deeply. "No dragon has claimed you as a pet," the nox-wing said. "Otherwise, you would be marked with its scent." He wrinkled his nose. "You reek of human. And something else. Something unappetizing. Perhaps you won't make a good snack after all."

"Oh, I would be a terrible snack," said Billy. "I'm mostly bones."

"Well, I do like something with a bit of crunch," mused the copper-colored dragon.

Billy chanced a look over the dragon's shoulder, trying to find his friends. They needed to get out of here. They needed to get somewhere safe. But he couldn't see them anywhere. The Dragon Court was enormous, and there were so many humans and dragons crammed in there, it was impossible to see everyone. He could only hope that they'd snuck away while they still could. But he knew in his heart they wouldn't leave without him, just as he wouldn't leave them.

Then the Dragon of Death spoke again, her cold voice reaching into Billy's very bones.

"The tournament begins tomorrow. Because I am a benevolent ruler," she paused as if daring anyone to contradict her, "I will grant permission to any human who wishes to enter the tournament. There are no rules except to survive and to win." She bared her teeth in a gruesome smile. "And the winners will, of course, be bonded with my most trusted dragons. The losers will be sent straight to the farms, as they are worth more in life force." Then she flung her enormous wings wide open. "I grow tired of being so low to the ground, like a snake or a worm. Remember, even in death, dragons belong in the sky. The earth, and all on it, is for us to consume. All the energy, life, and power is ours for the taking." She stretched out her long neck and opened her mouth. "Before I return to the top of the Tower, I will remind you all why I am your ruler. Why Dragon City is mine. Why you must all bow down—human and dragons alike."

As the Dragon of Death continued to speak about her great achievements and why all in Dragon City must worship her, Billy heard a scuffling next to him and very slowly glanced over. Ling-Fei, Charlotte, and Dylan were shuffling toward him. Billy breathed a sigh of relief that they were still together and alive and that they had found him. In the pandemonium of the Dragon Death Drop, Ling-Fei had left her perch on Charlotte's shoulders. Remarkably, Dylan still wore his hat and beard.

"Run," Billy mouthed. They had to get out of here while they still could.

His friends shook their heads. He remembered what they'd all promised each other. That they'd stay together. That they were stronger together.

He couldn't let them sacrifice themselves for him.

"Go!" he mouthed. There was nothing they could do for him right now. He needed them to know that.

Charlotte stared hard at him, then her mouth set in a grim line, and she nodded once. She pulled on Dylan's arm. He shook his head, but Charlotte yanked harder. Ling-Fei suddenly snapped her head up, staring toward the Dragon of Death. Something had caught her attention. Billy saw true fear fall over her face before she looked back at him. Then, very deliberately, she touched her earlobe, her heart, and her ear again. Billy instantly understood what she was trying to tell him.

Listen to your heart.

Listen to the heart bond.

And then Ling-Fei, Charlotte, and Dylan disappeared into the crowd. Away from the Dragon Court. Away from the Dragon of Death.

Even though it was what Billy had wanted, he still felt a rising panic. Now he was on his own. And then a thought entered his mind unbidden.

Close your eyes. Hold your breath.

Billy snapped his eyes shut and inhaled the biggest breath he could. Seconds later, there was a blast of power from where the Dragon of Death stood. Her voice echoed.

"It is my right to take energy from you all! Dragons and humans. All bow before me!"

Still keeping his eyes closed, Billy threw himself onto the ground in a bow. Brilliant white light shone so brightly that Billy could see it even with his eyes closed. He clenched his eyelids even more tightly shut, trying to block it out. And then the screaming started.

CHAPTER 10
FINGERS AND TOES

"This is just a taste!" roared the Dragon of Death. "A reminder that I can take life force from all of you whenever I wish! You fear the farms, you fear the factories, but all you truly must fear is me! And I will live forever! I will rule forever!"

A wave of pain rushed over Billy, and he felt like his very essence was draining out of him. It was the most terrible thing he'd ever experienced. It had to stop, it had to stop. He wanted to scream, but then he remembered the voice.

Hold your breath.

So Billy kept his mouth shut and didn't breathe in, but the noxious power was too strong, and it still seeped inside of him. He wasn't sure how much longer he could hold his breath. He was going to pass out. He was going to die.

And then, as suddenly as it had started, the pain stopped. Billy took in a gasping breath, expecting to breathe in the noxious cloud, to feel it fill his lungs, but there was nothing.

Very cautiously, Billy opened one eye and gasped again.

All around him, humans and dragons lay on the ground, shivering. Their mouths were gaped open, and their eyes were wide, almost like they couldn't shut them. Billy remembered the silent instruction that had echoed in his mind. *Close your eyes. Hold your breath.* His gaze shot to Spark, standing silently next to the Dragon of Death. She and the other dragons who stood on the huge platform overlooking the Dragon Court, the dragons closest to the Dragon of Death and their heart-bonded humans, were the only ones still upright. Could she have been the one to warn him?

Spark? Can you hear me?

He watched for any response. But just like the moment when she had swooped down on the thief, there was nothing.

Their heart bond must be broken. That was the only explanation. That was why Spark was going to get a new human to heart-bond with. As if he had needed any more proof. But surely he would have felt it, felt their bond break, just as he would feel if one of his own bones broke.

You have bigger things to worry about now, he told himself sternly.

Just then, another familiar face on the platform caught his eye. While it was clear that JJ hadn't had his own life force taken, he was pale, and he looked nervous. Next to him, his grandfather, Old Gold, gleefully cackled with the Dragon of Death. Old Gold looked different in this time. His white hair and white beard had grown down to his feet, billowing out around him in an unnatural wind, and his eyes shone with a strange brightness that Billy could see from here. The Dragon of Death appeared more terrifying than she had moments ago; her horns were even longer and sharper and she glowed with energy and power as she gazed out over the devastation and destruction she had inflicted on her own subjects. Even dragons weren't safe from the Dragon of Death and were at the mercy of her whims and wishes. She wanted all her subjects, humans and dragons alike, to fear and worship her. Very slowly and carefully, Billy stepped behind the copper-colored nox-wing to avoid the Dragon of Death's gaze.

"Dragons above all!" she shouted, and with a mighty flap of her wings, she shot to the top of the Tower. Old Gold zoomed up after her, holding his staff and riding a cloud that he summoned. Even though he was human, he was the Dragon of Death's heart-bonded human, and she was stronger with him by her side. A few moments later, Spark followed Old Gold and the Dragon of Death. As her shadow stretched out behind her, dragons and humans alike cowered. Despite everything, Billy couldn't bring himself to call her Death's Shadow, even to himself. She would always be Spark to him. No matter what.

The only dragons left on the platform were Da Huo and a shining white and silver dragon Billy had never seen before. JJ leaped up onto Da Huo's back. Billy still couldn't quite comprehend how JJ had become one of the highest-ranking humans in all of Dragon City. Surely JJ wasn't evil. Not like his grandfather, Old Gold. Not like the Dragon of Death.

As JJ and Da Huo stared out across the enormous courtyard of dragons and humans, all still trembling from having some of their life force taken, Billy thought he glimpsed a hint of remorse in both of their eyes, even in Da Huo's fiery ones. Fiery eyes that suddenly landed on Billy. The orange dragon startled, rearing his head back as confusion danced across his face, and then JJ's gaze followed. Billy could only hope that his disguise was enough to fool them. Surely JJ and Da Huo wouldn't recognize him without his hair and with the fake moustache plastered to his face and the long cloak draped around him. He turned his face away and found himself staring directly into the copper-colored nox-wing's red eyes.

The nox-wing let out a low groan from deep in its throat. "Now I know what I will use you for. I need more life force to make up for what the Great Dragon of Death has taken. It is her right, of course, as it is my right to take life force from you."

A shudder of terror ran through Billy, and he felt cold all over. He thought the worst thing had already happened today—being scooped up by this dragon and dropped from the sky—but it looked like things were about to get much, much worse.

Before he could formulate an escape plan, or even another thought, the copper-colored nox-wing threw him up on its back. "Hold on," it growled. "You are worth nothing to me dead."

"At least we agree on something," Billy muttered, but he held on as best as he could.

Riding a dragon that he wasn't heart-bonded with was much more difficult than riding Spark. Especially when it was a nox-wing who cut through the sky as if it were daring Billy to fall off. The wind stung Billy's eyes, but he forced himself to keep them open. He had to know where the nox-wing was taking him. Otherwise, he'd never be able to escape and get back to his friends.

A terrible thought occurred to him. Without him, his friends wouldn't be able to get back inside their safe haven underground. They needed him to travel through the life veins. Billy desperately hoped they could find somewhere safe to stay for the night, somewhere he would be able to find them. It was too dangerous to stay out in Dragon City after dark. And he hoped they wouldn't do something silly, like try to save him from the nox-wing, even though he knew if any of them had been taken, he'd be leading the charge on a rescue mission. But this wouldn't be a rescue mission—it would be suicide.

The copper-colored nox-wing took a sharp turn toward one of the diamond-shaped buildings, perched high on a long stem radiating with electricity. As the dragon approached, it let out a roar and a slat of the diamond opened up like a window so they could fly in.

Inside, everything was gold and shiny. The copper-colored nox-wing sighed with contentment as it unceremoniously dumped Billy in a corner, before it dived nose-first into a pile of jewels. Billy almost rolled his eyes. Dragons and their hoards.

But while the nox-wing was momentarily distracted, it was a good time for Billy to get his bearings. The diamond room had a huge, vaulted ceiling and the ground crackled with electricity. Billy dared a glance outside the still open window and shuddered. There was no way to get down unless the nox-wing flew him back to the ground. He was more trapped than Rapunzel in her tower.

He frowned as he gazed out the window. Something was winging its way toward them . . . and fast. A hot-pink dragon, as bright as a blazing sunrise, with a long white mane streaming out behind it, was heading straight for the diamond room. Straight for Billy!

Billy staggered back just in time as the dragon barreled into the room. Over half of its length was its tail, which had a sharp spike at the end. As it hit the floor, it sent up a cascade of sparks in its wake.

"You brought us a human!" the pink nox-wing crowed, licking its lips with a long tongue. It had short legs and scampered over the ground like a beetle, its claws clicking menacingly on the floor. Billy backed up until he was against the wall and had nowhere to go. He closed his eyes as the dragon's long prickly tongue patted him on his cheeks. "What a strange human!" the pink dragon went on, inhaling deeply through its protruding snout. Its nostrils flared as it did. "It smells so . . . young. But it has hair on its face! Does that not mean it is old?"

"I did not take the time to ask it its birthday," snapped the copper-colored dragon. "And it is not our human. It is *my* human. I won it in the Dragon Death Drop."

The pink dragon whipped its head around so fast that its mane whacked Billy in the face. "You know I could not attend the celebration because I was on duty in the dungeon today!" It blew out a huff of smoke from its nostrils. "We had a deal! I would take the shift during the anniversary, and we would share any spoils from the celebration!"

The copper-colored dragon snarled. "I do not want to share!"

Billy gulped. These nox-wings were talking about sharing his life force. He took a very small step back.

"You will share! Not only did I miss the celebration, and you know how much I love a party, I had to guard the top three today and they are such hard work! You cannot let them out of your sight for a second. The fat green one is always singing and sneakily healing the others. The giant red one blows smoke every time anyone gets near it. And the silver one looks like it would kill us all, given half the chance." The pink dragon spat on the floor. "I do not know why the Great One does not kill them and be done with it."

"Because she uses them as part of her personal life force source," said the copper-colored dragon. "They are worth more in life force."

"Tell that to someone who does not have to guard them all day every day," said the pink dragon with a humph.

Billy's mind was whirling. A fat green dragon? A giant red dragon? A silver one who looked fierce? That sounded like Buttons, Tank, and Xing! His friends had all known their dragons were alive because of their bonds, but they hadn't been able to track them down in Dragon City. And now he knew where they were! In the Tower dungeon. Before this moment, Billy hadn't even known there was a dungeon.

"I do not know why you are so impressed with yourself for winning *this* human as a prize," the pink dragon went on. "I could swallow it in one bite."

"Which is why I am not going to eat it. I am going to use it as my own life force source." The copper-colored dragon shuddered. "The Great One took life force from all gathered today. It is her right, but I am now in need of a recharge."

The pink dragon whipped its head around. "She took life force from the nox-wings?"

The copper-colored dragon nodded solemnly. "Do not look so surprised. She is the Dragon of Death. She did it to show her power. I bow down to her, today and always."

The pink dragon frowned and shuffled back and forth on its short legs. "But nox-wings are on her side. She has so much life force already. Why would she weaken her loyal ones?"

"Quiet your tongue!" hissed the copper-colored dragon. "Do not speak like that. Are you nox or not?"

"I am a dungeon guard. Of course I am nox!" the pink dragon shot out a warning streak of bright pink flame toward the copper-colored dragon, scorching the floor.

The copper-colored nox-wing glared back at the pink dragon, and, for a brief moment, Billy thought they were going to battle. But then the copper-colored dragon relented. "You are nox," it said. "But that does not mean I must share my human with you!"

"Give me the fingers at least. You know I like the fingers the best. I love how they crunch!"

Billy balled his hands into fists, hiding his fingers from view. No dragon was going to eat them!

"The vermin does not need its fingers to provide life force," the pink dragon went on. "Or its toes. Oh, I want its toes!"

Billy gulped. He had to get out of here fast.

"I want the toes," said the copper-colored dragon stubbornly. "But maybe I will give you . . . How many toes do humans have again?"

"Ten," said the pink dragon. "Ten toes and ten fingers—I want them all. And I want a sip of its life force every day."

"A sip of life force! I know how you guzzle it down." The copper-colored nox-wing shook its head. "No."

"YES!" roared the pink dragon, slamming its tail on the ground. The diamond room shook from the force of it.

As the nox-wings continued to bicker, Billy looked out the window and eyed the electric stem shooting up into the diamond that the dragons called home. How was the stem supporting such a huge structure? It was like a diamond flower high up in the sky. Billy wondered if the electric stem worked like the life veins and whether he could leap inside of it the way he could underground.

But it was too risky to attempt. If he was wrong, he'd fall to his death. In frustration, he jammed his hands in his pockets and suddenly felt a strange jolt. *That's odd*, he thought. There was something in his pocket. Something . . . buzzing with energy.

Making sure the dragons weren't watching him, he carefully took the item out of his pocket to examine it. It was the broken nox-ring. But the two pieces were buzzing in a way they hadn't before. He thought about how he could travel through the life veins, how his connection to Spark and maybe even the faint echo of his Lightning Pearl power had given him his own affinity for the electricity that fueled Dragon City. Billy gingerly pressed the broken pieces together, closed his eyes, and focused. He felt a whizz of power jolt from his heart, down his arm, and directly into his hand where he held the broken nox-ring. But when he opened his eyes, he saw it wasn't broken any longer. The ring, now the size of a baseball, glowed with electricity. He tentatively stuck his hand through it, and it locked onto his wrist, buzzing with power.

But he only had one nox-ring and he wasn't even sure how it worked. Would one be enough to carry him through the sky? Or at least enough to keep him from going splat on the ground? It was a better death than having his life force drained by these dragons.

He didn't have much time. The nox-wings were still bickering, but soon they would tire of that and want to devour him one way or another. He had to go now. The nox-ring was growing hotter and hotter around his wrist, as if it were leveling up in power and telling him it was now or never.

"Please work," he whispered, and then before he lost his nerve, he crept closer to the open window at the edge of the diamond. The wind whipped past him, pushing him backward. He heard the nox-wings still arguing. "I won it," said the copper-colored one petulantly. "You find your own human! I do not want to share its life force and that is that."

"If you do not want to share its life force than you'd better find somewhere else to live," snapped the pink dragon.

"Nobody is stealing my life force," Billy whispered to the wind. And with a desperate hope that the bickering nox-wings wouldn't notice his absence until he was long gone, Billy flung his fist out in front of him and leaped into the sky.

CHAPTER 11
THE THUNDER CLAN

Billy fell through the sky like a stone.

He had to do everything in his power not to scream and alert the nox-wings. Gritting his teeth, he forced his arm with the nox-ring above his head, hoping that it had enough power in it to at least slow him down, even if it wasn't enough to keep him aloft in the air. "Come on," he muttered. "Work!"

He felt a shock of power as the nox-ring tightened and flared, and then suddenly the wind wasn't as biting. He was still falling but more slowly now, as if he were buoyed by a parachute. A hysterical laugh escaped him. It had worked!

But now he needed to speed up. He had to get to the ground before the nox-wings in the diamond house realized that he was gone, or before any other dragons spotted him. The sun was already low in the sky and soon it would be sunset. He needed to land, find his friends, and get to safety—all before dark. Using every muscle in his body, he pushed the arm with the nox-ring downward and yelped as it immediately propelled him toward the ground. It felt like hanging onto a small and very powerful rocket.

But now Billy knew how to control the nox-ring and maneuver his way through the sky. No wonder the nox-hands were so desperate to earn a nox-ring—being able to fly, without a dragon, was amazing.

Billy remembered what had happened to the thief who had stolen the nox-rings and how Death's Shadow had come to take him away. Would Spark come for Billy? Would she catch him in her electric net and haul him to the Dragon of Death's Tower?

Spark is gone, he told himself firmly. And right now, he needed to focus on not smashing into the rapidly approaching ground. He adjusted his position again and aimed his feet toward a small alley. It was tucked behind a huge rectangular building that had purple smoke billowing out of a series of spiral chimneys splayed across the roof. As he descended through the smoke, he held his breath, careful not to breathe it in. But still he smelled it—the terrible and familiar smell of life force being drawn up from the belly of the earth itself.

Dragon City was all that was left. There was no energy, no life, anywhere else. Only the Void. Billy's feet hit the ground hard, and he collapsed to his knees. But he kept the arm with the nox-ring aloft, and as soon as he could stand, he yanked it off his wrist and hid it back in his pocket. Right now, he had to find his friends.

Billy racked his brain, trying to figure out where they might have gone. Where would he have gone if he had been in their place? They couldn't have made it through the life veins back to their train home without him. And he didn't know of any safe places for humans in Dragon City. They didn't have anywhere they could turn, anyone they could go to.

Except . . . Midnight.

Midnight knew their secret. Maybe Midnight would have snuck his friends into the orb where she lived and let them stay. It was huge in there; surely they could hide out from her father for a few hours, or maybe even a few days. And if they hadn't gone to Midnight, maybe Midnight would help him find them.

Billy quickly looked up, trying to orient himself. Even though he was in a part of Dragon City where he had never been before, he knew there was a way he could locate Midnight's home.

As always, the Tower stretched toward the sky, high above any other building. The clouds above it swirled and the entire structure crackled with energy. Billy knew if he could make his way to the Tower, he could figure out how to find Midnight's marble orb. And, hopefully, find his friends.

A screech from above distracted him from his thoughts. The copper-colored nox-wing was circling its diamond lair, bellowing in frustration. Even from down on the ground, Billy could make out what he was saying.

"Where are you hiding, vermin? You cannot have gone far!"

"Oh, stop pretending! You hid the human yourself! So you would not have to share it!" roared the pink one from its perch on top of the diamond.

With a gulp, Billy turned and ran as fast as he could toward the Tower. As he made his way through the winding streets of Dragon City, humans stared and most backed away as he passed. A running human was never without a dragon in pursuit, and nobody wanted to get caught in between.

Billy took a tight turn down a narrow alley and nearly jumped out of his skin when a hand reached out to try to grab him. A deep voice followed from the shadows "Wotcha running from, boy?" In the dim light, Billy saw a nox-knife glowing. "Did ya get yourself in some trouble? Maybe I can help you out. For a price, of course . . ."

The nox-hand's voice faded into the distance as Billy dodged past him, picking up the pace. Without thinking, he leaped up over the side of a building and then across to another alley.

Whoa. That was weird. That was the kind of thing Billy had been able to do with his pearl. Maybe he still did have remnants of his power . . . But no time to stop and think about it. He had to get to Midnight's home.

His legs were burning from running so fast and his lungs felt as if they might burst, but he pushed himself even harder. He hadn't survived the Dragon Death Drop and escaped those nox-wings for nothing. He needed to find his friends. Then a slow smile spread across his sweaty face.

There, at the top of the road, sat the giant marble orb where Midnight and her father lived. It was almost entirely round, and it appeared impenetrable. But Billy knew from experience that if you knocked on the right place, and the home wanted to allow you in, a hidden door the exact height of the visitor would slide silently open. As Billy approached the marble orb, he wondered why the Thunder Clan had chosen a home so close to the ground. The dragons in Dragon City aspired to be high in the sky at all times. Whatever the reason, though, it was lucky for Billy that the marble orb did sit on the ground. If it had been in the sky, like the copper-colored nox-wing's diamond home, it would have made things much more difficult for him.

But still, Billy hesitated. He'd never come here before outside of scheduled grooming appointments. What if he was wrong? What if Midnight's father punished him? What if they took him straight back to the nox-wings he had just escaped? He had to trust Midnight, but her father on the other hand . . . well he couldn't worry about that now.

He remembered what Ling-Fei had signaled to him. At the time, he'd thought she'd meant trust the *heart bond*, but maybe she meant trust your *heart*, trust yourself. And something deep inside of Billy told him that this was the right thing to do—that Midnight could be trusted.

You thought you could trust Spark too, he thought. For a moment, Billy was so overwhelmed with sadness, betrayal, and disappointment that he thought he was going to fall over and never get back up. That he would turn into a statue right there on the street—a warning to all humans that they could never trust a dragon.

He remembered what Xing had told him. *Do not harden your heart*. That conversation now felt like a lifetime ago. She hadn't been speaking of Spark's betrayal, she hadn't foreseen it, because as fate would have it, Spark herself was the one who was a seer, but Xing's words rang true. And Billy clung to them now.

He couldn't let Spark's betrayal make him grow wary of all dragons. He knew good dragons existed. Just as he knew, deep in his bones, that goodness could win out. Goodness *would* win. Spark had betrayed him, but Spark was not here. He had to trust his instincts. Trust his own heart. Spark had gone nox, not him.

Steeling himself, he took a deep breath and knocked on the marble, exactly where he'd knocked dozens of times before when he'd come to groom Midnight and her father. He waited, his whole body tense with anticipation.

Nothing.

Billy sighed with disappointment. But he wasn't going to give up. He knocked again and this time pressed his face right up against the marble.

"Please let me in. I'm a friend, I promise."

A few moments later, the marble slid open, and Billy jumped in before it could change its mind. The marble sealed itself closed behind him, and he desperately hoped he'd made the right choice by coming in. If he was wrong and Midnight's father wasn't happy to see him, well, now he was trapped inside the marble orb with no way to escape. He'd have to face Midnight's father and hope that he could convince the huge dragon not to turn him in.

Billy stood in a cavernous room, one where he'd been before. But this time it was empty. "Hello?" he called out tentatively. "Is anyone home?" His voice echoed against the rounded ceilings. "Midnight?"

As he looked around, Billy noticed a small light flickering down one of the enormous corridors that led deeper into the orb. A small light that was growing closer, and closer . . .

And *squeaking*.

It was the tiny gold flying pig! Billy ran toward her, his feet slapping loudly on the marble floors beneath him. If the pig was here, his friends couldn't be far.

Then a voice rang out from farther within the orb. "Slow down, little piggy! Where do you think you're going? I don't want you getting lost in here." Dylan turned the corner and came to a sudden stop, his jaw dropping open. Then his face split into a huge smile.

"Hey, guys! Billy is alive! And he's here!"

Billy laughed as he launched himself at his friend. "Dude! You're here!"

Dylan hugged him tightly. "Dude!" he said, imitating Billy's American accent. "*You're* here!"

"We're all here!" cried Charlotte, running into the entrance hall and joining the hug. Ling-Fei was right behind her, cheeks pink with excitement.

"Billy!" she exclaimed, beaming as she patted his shoulder. "I knew you'd find us."

"She's right, she kept telling us we had to stay here because this is where you'd come looking for us," said Charlotte.

"I thought you were a goner for sure!" said Dylan.

Charlotte whacked Dylan on the arm. "Don't listen to him. We knew you'd be able to escape that nasty nox-wing."

"And we were going to come up with a plan to rescue you!" added Ling-Fei, eyes bright. "But you didn't even need us!"

"How did you escape? And how did you know where to find us?" said Dylan.

Billy grinned at his friends and pulled out the nox-ring from his pocket. "This thing helped me out," he said, still in awe. "I was able to repair it somehow, using power that I didn't know I still had. I think it's related to my ability to travel in the life veins. Once I had the nox-ring working, well, all I had to do was jump out of the dragon's lair and then find you guys. And this is where I would have come if I were you. Clearly, it was the best place to go!"

Dylan shook his head adamantly. "Oh, no, you can't take credit for a plan that you weren't even here for!"

"What can I say? It was a great plan." Billy took a deep breath. "Seriously, though, I'm so glad to see you guys." He peered down one of the corridors. "Where's Midnight? And what does her dad think about us being here? How did you even get into the orb?"

"It was the strangest thing," said Ling-Fei. "The orb let us in on our own. It was like it recognized us. Midnight and Thunder were already here, and as soon as we came in, Midnight leaped in front of us to protect us."

"She did?" said Billy, a warm feeling spreading throughout him as he thought of the small dragon.

"It turns out there was no need for her to do that," said Dylan. "Her dad has been pretending to be a nox-wing, but I appreciated it all the same!"

"Why is he pretending to be a nox-wing?" said Billy.

"We've got lots to tell you," said Charlotte, rocking back and forth on the balls of her feet. "Everything isn't as it seems in Dragon City."

Billy scrunched his face up in confusion. "What do you mean?"

His three friends exchanged a look. Then Ling-Fei turned to Billy. "Billy, it's time for you to officially meet the Thunder Clan."

Billy followed his friends down one of the many corridors snaking out from the entrance hall of the marble orb. Even though he'd been here many times before, the dragons usually met him and his friends in the great hall to be groomed. He couldn't believe how many giant corridors and other rooms there were in the marble orb.

Finally, they reached a room that Billy guessed had to be the center of it all. The beating heart of the orb.

Inside, Billy wasn't surprised to see piles of jewels and gold on the ground and crackling electricity climbing the walls. It made sense that this clan would guard their hoard in the safest part of their home. Midnight and the black and silver dragon that Billy knew was her father were waiting.

But then there was a low rumble from the pile of jewels. Billy looked more closely at it and gasped. Those weren't just jewels on the ground, it was a jewel-covered dragon, each scale a different color, emitting its own light and sparkle. The dragon was fast asleep and gently snoring. With each exhale, more electricity snaked up the walls of the round room. But instead of flowing outward and onward, as all power in Dragon City did, it looped back around, staying in the room.

"You aren't the only one with secrets," said Midnight.

Billy's jaw dropped. The black and silver dragon let out a booming laugh at his expression. "Welcome to the Thunder Clan."

CHAPTER 12
A TRADING OF TALES

The giant sleeping dragon continued to snore. Billy stared at it in awe. Even after seeing so many dragons, flying on them, and battling them, there was something special about this one. With every intake of breath, electricity traveled up to the ceiling and then spiraled back down and into the dragon itself, giving it a faint pulsing glow.

"This is my mother," Midnight said shyly. "She's like Death's Shadow . . ."

"Like Spark," Charlotte interrupted, as if Billy could ever forget that the hated Death's Shadow and Spark were one and the same. But then he realized that if Charlotte was using Spark's name, her real name, they must have told Midnight more about their history with dragons.

"Go on," Billy said, his voice suddenly hoarse with emotion.

"She's a lightning dragon. Incredibly rare and very powerful. She generates more electricity and power in one breath than most dragons could create in a day." Billy couldn't miss the pride in Midnight's voice.

"That is why we have to keep her hidden . . . and asleep. If the Dragon of Death knew we had a lightning dragon, she would take her. At best, she would make her into another Death's Shadow; at worst, she would drain her for all she is worth," added Midnight's father.

Billy stared at him. "And what kind of dragon are you?"

The dragon chuckled, his long moustache and beard quivering as he did. He seemed like an entirely different dragon than the one Billy was used to grooming. He was still intimidating, of course—after all he was a giant dragon—but it felt as if he were on their side. And that was a wonderful feeling. "I am what is known as a thunder dragon. But I will not demonstrate now; I would not want to blow your eardrums out."

"Thunder and lightning," Dylan mused.

"We came across a dragon once who had such a loud screech, it could stun anyone for miles around. Even underwater," Charlotte said slowly. "Is your power like that?"

"It sounds like you are talking about the Screaming Serpent. But everyone knows that is just an ancient legend," scoffed the thunder dragon.

Billy and his friends quickly nodded. "Yes, a legend," said Ling-Fei.

"I am impressed with your dragon knowledge." The thunder dragon appraised them. "It is rare for humans to know so much about our kind, even though we live closer now than we ever have."

"What should we call you?" Billy blurted out. "I know that a true name comes from a heart-bonded human, but there must be something we can call you. Before now, I thought you were . . ." Billy paused. "I thought you were a nox-wing!"

"That is because I live as a nox-wing to protect myself and my family. I am sorry if I was ever harsh to you. I hate how the nox-wings treat humans, but I hope you can understand. And back to my name, you may use my dragon name. In your tongue, it translates to Thunder. Appropriate, is it not?"

"Thunder, Lightning, and Midnight," Billy murmured. "The Thunder Clan." He shook his head. "But . . . how long has Lightning been asleep?"

"For many, many moons," said Thunder, his voice deep with sorrow. "Things were never as bad as they are now. Long ago, humans and dragons lived in peace, alongside each other. But then the Dragon of Death's power grew and grew, and it seemed as if it almost happened overnight."

"Mmm," said Dylan meaningfully. "That is . . . interesting."

How could they even begin to explain to Thunder and Midnight what had happened? That this future was a flawed one, that it only existed because the Dragon of

Death had chosen her destiny from the Destiny Bringer himself.

As much as Billy hated this dark future, he couldn't imagine extinguishing it and everyone in it. There had to be a way that they could stop the Dragon of Death, save the future, and return safely home. But he couldn't let himself think that anything else was possible; otherwise, he might start to crumble and crack, and he knew he needed to be strong.

They were the only ones who knew how the Dragon of Death had grown to such terrifying power.

They were the only ones who could stop her.

But they couldn't do it alone. They needed help.

"I may have been pretending to be a nox-wing, but my heart is still good. And because of that, I can sense goodness in others. Even in humans. That is why I know we can trust you. We have shared our greatest secret with you, and, in return, we want you to trust us. We know there is a lot that you are not telling us," Thunder went on. "We cannot help each other if we are not honest."

Billy looked at his friends, asking a silent question with his eyes. They each nodded.

It was clear they had already told Thunder and Midnight some of their story, but now it was time to tell them everything.

Billy took a deep breath. "We're from another time."

It took hours to tell Thunder and Midnight everything. They had so many questions and they kept interrupting.

When Billy told them, haltingly, how Spark had betrayed them, how she had become Death's Shadow, Midnight came and nuzzled him gently.

"And that's the end of our tale," said Dylan, after he'd finished explaining about their disguises and attending the Five Thousand Year anniversary celebration.

"Well, not quite," said Billy. "I know where our dragons are."

"What? And you waited until now to tell us?" spluttered Charlotte.

"I thought we said no more secrets," said Ling-Fei, hurt dancing across her face.

Billy ran a hand through his hair. "I'm sorry," he said. "I meant to tell you right away, but I didn't know if we could trust Midnight and Thunder." He glanced up at them. "No offense."

"We understand," said Thunder. "But tell us, where are your dragons? They must be missing you as much as you are missing them."

Billy quickly told them about what he'd overheard while in the copper-colored nox-wing's diamond lair. "It has to be our dragons. They must be in the Tower dungeon."

Midnight gasped. "Dragons who enter the dungeon never come out."

"It is true," said Thunder. "Our greatest fear is that Lightning could be taken into the dungeon. That is why she must sleep—to keep her powers hidden from the Dragon of Death."

"We've saved our dragons before," said Charlotte stubbornly. "We can do it again!"

"The Tower dungeon is not like anything you have seen before," said Thunder gravely. "All prisoners are trapped behind the Flames of Death."

"Flames of Death, Death's Shadow, Dragon of Death . . . they should really call this place Death City," Dylan muttered.

"Once you pass through the Flames of Death, there is no return," said Thunder, looking closely at Billy. "Do you understand?"

"We have to try," said Billy. "It's our only hope."

"We cannot help you get into the dungeon," said Thunder. "I am sorry. We have to protect Lightning. We have to continue to pretend to be nox-wings. It is the only way we can stay alive."

"I understand," said Billy stiffly.

"But we promise to keep your secret. And we will offer safe haven for you here, whenever you need it. We just cannot go into a battle when we know we will surely lose."

"You've already done so much for us," said Ling-Fei kindly. "We appreciate it."

"So you're okay with letting Lightning sleep forever?" demanded Charlotte.

"Dragons can live a very long time," said Thunder. "She is merely taking a nap. The Dragon of Death may not rule forever."

"She will if nothing changes," said Charlotte, glaring

at Thunder. She turned to Billy. "We're going to save our dragons, aren't we?"

"Of course we are," said Billy.

"We should have known they'd be in a dungeon," said Dylan. "Of course the Tower has a dungeon. That's exactly something the Dragon of Death would love. It seems so obvious now!"

"If you are intent on going to the dungeon, at least wear these cloaks." Thunder nodded at Midnight, who flew off down another corridor and quickly returned, gripping four shimmering cloaks in her claws. The fabric mirrored the shining, multicolored scales on Lightning, changing colors in different lights. It reminded Billy of when the sun hit an oil slick, turning it into a rainbow.

"That's dragon fabric," said Billy, recognizing it as the same fabric their old dragon suits had been made of.

"Lightning and I heart-bonded with humans, long ago. We made these cloaks for them. But since they died, we have had no use for them . . . until now."

Dylan swallowed audibly. "You want us to wear cloaks that belonged to dead people?"

"Shh!" said Ling-Fei. She took the cloaks in her arms. "Thank you for your generosity."

"These are actually surprisingly stylish!" said Charlotte, fastening her cloak around her neck. "And if the original owner is dead, well, that just means it's vintage."

"Father, I want to go with them!" exclaimed Midnight. "I can help!"

"Absolutely not," said Thunder sternly. "You are staying here with me and your mother. I still need to talk to you about the fact that you snuck into the life veins! How were you even able to do that?"

Midnight nodded at Billy. "He was able to do it!"

"Oh, that's the other thing," said Billy, fumbling in his pocket. He took out the glowing nox-ring and held it out. "The broken pieces of the nox-ring aren't broken anymore!"

"Interesting," said Thunder, studying it. He gave Billy a long look. "Something in you fixed it."

Billy swallowed. "I can tell I have some of my old powers, but I'm not sure how." He hoped that Thunder would reassure him that it had nothing to do with Spark, but then the big dragon spoke.

"The heart bond goes deep. Perhaps you have some of your dragon's electric power."

"I don't want anything to do with Spark!" Billy threw the nox-ring on the floor, where it shattered in two and immediately lost its glow. "I'm not like her! I'm not nox!"

"It sounds like she was not always nox," said Thunder, before carefully picking up the broken nox-ring pieces to examine them. "Fascinating. You know, I have never seen a nox-ring up close."

"Billy, I think Thunder has a point," said Charlotte. "It isn't only our heart bond that's strong. I still have some of my super strength. Not as much, of course, but I can tell it hasn't completely disappeared."

"Same for me," added Ling-Fei. "I can still sense magic and life." She let out a light laugh. "I don't think

I could open up the earth again, but it's nice to feel a flicker of it."

"I haven't tried my power yet," said Dylan. "With my luck, it probably doesn't work at all anymore."

"You say your power was charm?" asked Thunder. Dylan nodded. Thunder raised a furry eyebrow. "And you wonder how you survived the anniversary celebration. Your disguises were not that good, I must tell you."

"You mean, I charmed dragons and other humans into believing our disguises?" Dylan let out a delighted laugh. "Good to know I'm not entirely useless!"

"And I definitely still have . . . *something* in me," said Billy. "It's not just my agility, either. I feel so much more connected to the life force powering Dragon City, and the electricity too." He turned to Thunder. "We're better prepared for this than you think we are, but we could still really use your help."

Thunder shook his mighty head. "I must protect my clan, small as it is. But I wish you luck."

"How are you even going to get into the Tower?" said Midnight. "It's guarded at all times. Any human found there will be taken straight to the farms, or worse, used as the Dragon of Death's personal energy source!"

"Except for tomorrow," said Billy, an idea solidifying in his brain like water turning to ice. "Because tomorrow the tournament begins, and everyone will be distracted. It's the perfect chance for us to sneak into the Tower."

"You're right! The whole place will be crawling with humans!" Dylan held out his hand for a high five.

"The more humans there are, the more nox-wings there will be. It will still be incredibly risky," said Thunder.

"It'll be dangerous, but I think it's our best chance at finding our dragons," said Ling-Fei.

"We've never let danger stop us before," said Charlotte. She grinned. "And I do love a tournament!"

"Remember, Charlotte, we don't want to actually enter the tournament. We want to go unnoticed," said Billy, but he was grinning too.

Charlotte tossed her long hair over her shoulder. "It's hard not to notice me, but I'll try my best."

CHAPTER 13
TO THE TOWER

The next day, Billy and his friends made their way through the streets of Dragon City back to the Dragon Court. They had spent the night in their train carriage, happy to be reunited in a place where they felt safe, and they were hopeful that they might soon be reunited with their dragons too.

The plan would work, Billy told himself as they walked. The tournament would give them the distraction they needed to sneak into the Tower and once they found their dragons, they could do anything together, even if Spark wasn't on their side. A pang of loss pierced Billy's chest. He still couldn't believe Spark had become Death's Shadow. He shook his head, trying to fling the thought from his mind, but the feeling hung on him like a heavy curtain.

Billy saw a flash of gold flicker in front of him and something nuzzled his cheek. It was the tiny flying pig. It was almost as if she had known Billy needed the support. He held out his hand and the pig flew onto his palm. "Do you want to come to the grand tournament with us, Goldie?" He'd started using the nickname for the pig, and it felt right.

The little pig trotted in a small circle on Billy's open palm and nodded her head.

"Are you sure?" Billy asked. "It's going to be dangerous."

The pig seemed to contemplate this for a minute before stomping her feet and snorting her snout. She looked back up at Billy and snorted a few more times.

"Only if you get a snack first?"

The pig leaped up and down in Billy's hand in agreement.

He reached into his pocket and pulled out a Life Bar wrapper. "This is all I've got." He shrugged.

The pig snatched the wrapper away from Billy and gobbled it down in three quick bites. She let out a satisfied burp.

"It seems you'll eat just about anything," said Billy with a laugh.

With a flutter of her wings, the tiny gold pig zoomed up to nuzzle Billy again before zipping into the pocket of his tunic. She wriggled around, sinking farther into the fabric, until only her gold snout poked out.

"All right," said Billy, "You can come along. But feel free to fly away if things get scary." He gave the pig a gentle rub between her ears, and she responded with a satisfied snort. Billy grinned. It was nice to have the tiny gold pig with him.

As they stepped back into the Dragon Court, Billy saw that the grounds looked just as festive as they had the previous day. Even more dragons were gathered on the floating platforms above them. In front of them, at the center of the Court, lay an enormous slate made of sparkling rubies that spilled out across the grounds like a pool of blood.

Charlotte shivered next to Billy. "There's something about that giant shining red floor that's giving me the creeps."

Ling-Fei nodded. "Yes, I can sense something very sinister about it. It almost looks like a battleground."

Billy could see that a crowd was slowly gathering around the rubies but stopped short of stepping onto it.

"That must be where the tournament is starting," said Ling-Fei.

"Well, in that case," said Dylan, "I say we stay as far away from that red battleground as possible."

Billy looked around and realized that they were the only children in the crowd, and he could tell by the tattered clothes of the humans on hand that only the most desperate were present. A wave of doubt struck Billy. Were they making a terrible mistake by coming to the tournament? He pushed the thought out of his head. Dylan, Charlotte, and Ling Fei had to get their dragons back. And this was their best shot at getting into the Tower, which was on the other side of the ruby platform.

As Billy tried to figure out how they could sneak away from the crowd, a chill slithered into his bones, and he suddenly felt as if he might be sick. He looked up and saw the Dragon of Death high in the sky. She swooped down onto a podium above the ruby battleground, with Da Huo, JJ, Old Gold, and Death's Shadow following in tow.

"Loyal subjects! Welcome! I can see that many of you have accepted the generous offer I have made to allow the very best of you to bond with my most powerful dragons. Let me be clear, though, only the very best of you will succeed. This tournament will be the most challenging thing that any of you have ever attempted and only those who are truly worthy will be allowed to bond. But the bond will, of course, be worth it for those who succeed. The winners will have power beyond anything ever seen in a human." The Dragon of Death paused and looked up toward the sky. Confetti was raining down around them. She let out a smile. "Before we begin, there will be a dragon dance to celebrate this momentous occasion."

She roared a mighty roar as two long and slender dragons shot out from behind her and hovered over the ruby platform. They swirled round and round with the grace and elegance of smoke rising from a bonfire. Dragons all around them stomped their feet on the ground in unison and the earth shook in rhythm with the dance. It was mesmerizing.

And then a thought struck Billy. "Quick! This is our chance to get into the dungeon while everyone is distracted! Let's sneak around the crowd to the base of the Tower."

"I was just about to say the same thing," said Charlotte with a smile.

The four friends quickly pushed through the edges of the crowd, around the platform where the dragons were dancing and past the podium toward the Tower. The onlookers were so entranced that they hardly noticed the children.

As they reached the back edge of the crowd, the four friends could clearly see the base of the Tower in front of them. They were about three hundred feet away.

"What do we do now?" asked Dylan. "If we want to get any closer to the Tower, we're going to have to break away from the crowd and then we'll be out in the open with no cover. We'll definitely be seen. And then caught. And then probably eaten."

"What if you tried to use your charm to hide us?" said Charlotte. "Remember what Thunder said—our disguises weren't very good so you must have helped us to stay hidden yesterday. I'll bet if you try, it'll be enough to cover us while we run to the base of the Tower." Charlotte looked around at the group. "What do y'all think? You in?"

"We'd better run really, really fast," said Ling-Fei.

"I think it's our best shot," said Billy.

"I don't know, guys," said Dylan, looking worried. "I don't think I can do it."

"You can!" said Billy, gripping Dylan's shoulders and pulling him close. "You've got this. We need you! Our dragons need you!"

Dylan let out a sigh. "Okay, I'll do my best. But if we get eaten, I am taking zero blame."

"We believe in you," said Ling-Fei. "Now hurry, the dance will probably be over soon. We don't have much more time."

Dylan took a deep breath and nodded. "Let's do this. Run as fast as you can toward the Tower on three. One. Two. Three!"

Billy felt a rush of adrenaline as he and his friends dashed into the open space between the crowd and the Tower, hands linked and running as fast as their legs would carry them.

"I think it's working," said Dylan, his voice strained. "I think I'm keeping us hidden!"

"Keep it up!" said Ling-Fei.

"You go, Dylan. I knew you could do it!" said Charlotte between breaths.

A ball of fire exploded in front of them, forcing Billy and his friends to stop in their tracks.

"You there," said a strange voice. Billy's heart sank. "What do you children think you're doing?" A nox-wing had spotted them.

Dylan whimpered. "I'm sorry, guys. I tried my best. I guess I'm not as charming as I used to be. And that fireball distracted me too."

Charlotte took a big step toward the nox-wing who was floating in midair. "What do you think we're doing? We're here for the tournament, of course."

The dragon narrowed its eyes. "You know very well that you are heading in the opposite direction from the tournament."

Billy nudged Dylan in the side. "Now's your chance."

Dylan shot Billy a confused look.

"Use your charm," Billy whispered quickly.

Dylan still didn't look very confident, but he turned back toward the dragon and took a step forward. "Oh, mighty nox-wing, yes, you're right. You've caught us. We're here for the tournament, but the Tower is just so magnificent, and we couldn't help but try to get a view of it up close. Please find it in your heart to forgive us."

The dragon let out a snort. "My heart has no room for forgiveness. The Tower is strictly off-limits, which you should very well know. But the Great One wants as many humans as possible to compete, so you may return to the tournament. Just know that I am watching you. Make one more suspicious move and I will not let you go again."

The four friends were herded back to the tournament just as the dance was finishing. The Dragon of Death let out another mighty roar. "Now it is time. All those who wish to take part in the tournament, come forward and stand on the ruby ground that I have laid before you."

The crowd shifted as the humans considered their options. Billy could feel the eyes of the nox-wing on them as he looked at his friends. "I think we have to join the tournament. It's our best shot at finding an opportunity to sneak into the Tower and I don't want to find out what the nox-wing will do to us if it finds out we were lying."

He exchanged a look with his friends, who only nodded. Without speaking, they all turned and stepped onto the ruby battleground.

CHAPTER 14
THE TOURNAMENT

As they settled onto the ruby floor, an older boy with dark curly hair and deep brown eyes caught Billy's attention. "Are you scared?" he said, gazing at Billy. "I'm not scared," he went on, not waiting for Billy's reply. "My mom says I was born for this. I was born to be bonded with a dragon and to help our family move into a palace in the sky. She knew this day would come and I've been training for it since I can remember." Billy couldn't help but notice that the boy looked especially strong. He had a thick neck and muscular arms.

"We might live underground now," the boy continued, "but this is our chance to move up in the world. Remember the name Timothy," he said. "That name will soon mean something to everyone in this city." He flashed Billy a wide grin before turning back toward the Dragon of Death.

"Time is up!" she roared. "Every human on the ruby platform is now officially a contestant in the tournament. The tournament will consist of two stages. Stage one will be the elimination phase, where we will remove contestants until only twenty of you are left standing. Stage two will be when the winners are decided. The twenty remaining competitors will race through the Tower that stands before you. The first four humans to finish the race will earn the great privilege to bond with one of my dragons. And as I announced yesterday, the overall winner will bond with Death's Shadow herself!"

Billy felt like he'd been punched in the stomach. Even though he had heard it yesterday, he still couldn't accept that Spark was going to bond with another human. Blood rushed through his veins, but his chest felt empty. And then a thought struck him. What if he won the tournament? Would he want to be bonded with Spark again? Before Billy could contemplate it any further, the Dragon of Death came back into focus. "Here is your first challenge!" she roared.

A large circular table rose out of the center of the ruby battleground and resting on top of the table were rows and rows of ruby collars. "Snap the collars around your necks," spat the Dragon of Death.

The humans around them hesitated, looking to see who would go first and whether something would happen to them when they snapped the collar on. But no one moved.

"Now!" roared the Dragon of Death, firing a burst of flames above the crowd. It was so close to where Billy was standing, he thought that what was left of his hair might have caught fire.

The contestants shuffled forward now. Those who snapped the collar on first seemed to do so without consequence. Billy and his friends each grabbed a ruby collar and held it in their hands.

"Are we sure about this?" said Dylan.

"We've got each other," replied Ling-Fei. "And together, we can do anything."

Billy nodded and snapped the collar around his neck. The cool touch of the stone sent a chill down his spine. He gave his friends a weak smile. "So far, so good."

His friends returned the smile as they put on their collars, their eyes determined.

"Very good," said the Dragon of Death when everyone was ready. "The tournament begins now! Only those who can escape the collar will proceed to the next challenge."

The last thing Billy heard was the Dragon of Death's laugh as the collar tightened around his neck. He gagged and tried to pull it off, but the more he pulled, the tighter the collar became. He pulled and pulled and pulled, until the collar was so tight that he struggled to breathe. He turned toward his friends for help, but he was shocked to find that he was standing alone on the ruby platform. The crowd and all of the competitors and dragons had disappeared. It was just him. And he was completely alone. Billy sank to his knees and his vision blurred. He fell forward and then his vision went completely black before his head hit the ground. He felt as if he was being tipped from a cup as he tumbled into a black abyss.

Then, in flashes, his vision came back. But he wasn't in the Dragon Court. No, he was on a beach at home. And in front of him was Spark! And next to her were his parents and his brother! Spark looked over at Billy, and in horror, he saw her eyes were completely blacked out. No! Spark was going to betray him; she was going to hurt his family. He couldn't let her do it again. And then suddenly he was in the ocean, swimming after them. He had to reach them, he had to convince Spark to turn back to her old self. Back to being good. Wave after wave struck him in the face and Spark and his family grew smaller and smaller in the distance, but he had to save them; he couldn't give up. The fear of what would happen if he didn't reach them was so strong it almost dragged him down beneath the current.

Billy felt something grasp his hand and then he was tugged backward. The warmth from his hand spread up his arm and into his chest until his whole body felt as if it were being warmed by the sun. His vision grew dark again and then he blinked.

"Billy, are you okay?"

Billy blinked again and Ling-Fei came into view.

"Thank goodness you're awake, Billy!" Ling-Fei's eyes were red, and her face was covered in tears. "Hurry, we have to save Charlotte and Dylan. They still have their collars on!"

Billy got up, and as he did, his collar fell off with a thud next to him. He was back on the ruby platform, Charlotte and Dylan writhing on the ground next to him.

"Grab their hand and try to speak to them. I think it's driven by fear—the more afraid you are, the farther you get sucked in," said Ling-Fei. "And the tighter the collar gets."

Billy lunged over to Dylan as Ling-Fei grabbed Charlotte by the hand.

"Dylan!" Billy yelled, grabbing his hand. It was as cold as ice. "Dylan, it's me, Billy! You're okay, you're safe. Follow my voice."

Dylan's eyes were squeezed shut and he was still writhing on the floor but less so now. Billy was getting through to him.

"Dylan, follow my voice," he repeated. Dylan blinked, once, twice, and then let out a strangled cry.

"Billy? Where are we? I saw my sisters!" His collar fell off from his neck as he spoke, its spell broken.

"It was all a trick," Billy said grimly. "Part of the tournament. Your sisters aren't here. But you are. You need to focus. We have to stay alert."

Next to them, Charlotte sat up, looking dazed and rubbing her neck. "How . . . how did it do that?" she croaked.

"Dark magic," said Billy. "Luckily Ling-Fei was able to break through on her own and she woke me up. Then we were able to save you and Dylan."

All around them, competitors who had not escaped their nightmares fell to the ground as their collars tightened more and more. Nox-wings swooped in and took them away.

The Dragon of Death let out a loud, cruel laugh. "Well done to those of you who survived the first challenge. As you now know, it was a mental one. This next challenge will be physical, and it begins now!"

As the Dragon of Death spoke, electricity began to flow into the ruby stone beneath their feet. And with a deep *crack*, the stone began to undulate, knocking Billy and the group off-balance. The crowd started shifting back and forth with the ripples, and the more the crowd shifted around, the more the surface began to swell. Soon, huge ruby waves rose out of the ground like sharks, swift and silent. And when the waves crested and crashed, they swallowed everything in their path.

The Dragon of Death let out a sinister laugh. "Do not let the waves catch you or they will suck you straight to the farms! Get out alive and you will make it to the next stage of the tournament."

People were zigzagging in all directions now, trying to outrun the waves that moved like packs of predators, chasing down and swallowing those who stumbled or fell.

"Run!" said Billy. "We have to get out of here!"

Packs of waves were forming in every direction now. "This way!" cried Charlotte, leading them back toward the Tower.

The group of friends followed Charlotte as quickly as they could. Billy saw that the boy called Timothy was just ahead of them, jumping effortlessly off the waves that chased him. He dodged left and right, bursting away from the crests as they crashed around him. As he moved, more and more waves formed around him until he was surrounded on all sides like a mouse in a bear trap.

"No!" Billy yelled as he tried to speed up to save the boy, but he was too late. The waves crashed down on Timothy with a thunderous *boom*. And then the ruby surface went still again, leaving nothing behind.

"We have to go faster!" yelled Charlotte, her blonde hair whipping behind her.

Ling-Fei lunged forward and grabbed Charlotte's wrist. "No, wait! I think we should stay still! There's something strange about how these waves are moving."

Charlotte looked back at Ling-Fei as she continued to run, dragging her friend with her. A huge wave was closing in on them. "We'll get caught if we stop!" yelled Charlotte. "We have to keep running!"

"I think Charlotte's right," said Dylan. "We need to run!"

"Watch out!" cried Billy as a new wave rose right next to them. Billy dived at his friends, tackling them as the wave crashed down, barely missing them. He let out a sigh of relief before realizing that there was no way they were going to outrun the huge wave behind them.

"We're doomed!" cried Dylan, covering his face with his hands. In moments, the wave would devour them and suck them into the farms.

Ling-Fei reached out and grabbed her friends in a close huddle. "Shh!" she said. "Stay as still as you can and don't make a sound."

"I really hope you're right about this," said Dylan.

"Do you trust me?" asked Ling-Fei, her eyes wide and her face serious.

The friends stared back at Ling-Fei. They'd all been through so much together. They'd traveled through time and bonded with dragons and escaped death more times than Billy cared to count. Billy nodded back at Ling-Fei; of course he trusted her.

"Well, I don't think we've got much of a choice," said Dylan lightly, but his expression was serious.

"Come closer," said Ling-Fei, and they huddled together until they were so close their heads rested against one another.

"I really hope this works," said Billy. "But if it doesn't, there isn't anyone else I'd rather be going down with."

"This *will* work," snapped Charlotte. "But only if you stop talking! Now isn't the time for a motivational speech!"

"Shh," said Ling-Fei again gently.

As the wave approached, Billy felt the ground rise and tilt as if they were sitting on the rising side of a seesaw. Up and up they went, until the ground beneath them was so steep that they all tumbled forward.

"Keep still!" said Ling-Fei as they huddled closer, arms locked around each other. "I think it's working! A ruby wave hasn't crashed down on us yet."

Arms still intertwined, the four tumbled down the slope like a rolling pin.

"You're right. The waves aren't crashing onto us!" whispered Dylan in excitement. "It's like they don't know we're here if we don't move."

The wave chasing them began to slow down and shrink in size as if it were trying to figure out where they had gone. Another group of humans dashed past them, and the wave surged up in pursuit, leaving Billy and the others behind.

"Ling-Fei, you're a genius!" said Charlotte, hugging her. "We would have been ruby-wave toast if you hadn't saved us."

Ling-Fei smiled. "We make a good team. But we need to focus. Now that we know how to avoid the waves, we have to save as many people as we can."

"You're right," said Billy "You guys stay here. I've got a plan."

Billy ran toward the center of the battleground and cupped his hands around his mouth so he could yell as loud as possible. "Everyone! Stay as still as you can! These waves are attracted to your movement and sound!" As he shouted, the waves on the battleground stopped chasing the other contestants and raced toward him instead.

"Oh, no you don't, Billy Chan," yelled Charlotte as she too started shouting and waving her arms as she ran in the opposite direction.

The waves seemed to grow confused, pausing for a moment before they half shot toward Billy and half toward Charlotte.

"Over here!" cried Ling-Fei as she ran away from Billy and Charlotte. "Come and get me!"

"No get me!" said Dylan, joining in.

The waves were moving in circles now as they

switched between chasing the four children. The other contestants were getting away. Their plan was working.

Billy heard a piercing roar from above and the flapping of wings. The Dragon of Death was hovering over them. "What an impressive bunch you are. Who figured out my challenge so quickly? Your energy burns so deliciously bright." She smacked her lips. "And how foolish you are to try to save the other *competitors*. You do know that this is a competition, don't you?" She let out a cruel laugh. "I don't think I will ever understand you humans." She gazed at Billy with piercing eyes and then her eyes lit up in recognition. "*YOU*," she roared. "You filthy vermin! How dare you try to beat me at my own game?"

Her eyes lit up and the waves sank back into the ground, but in their place, a huge ruby ring rose around the four children, enclosing them like a lasso. And then the walls started to close in on them. Billy and his friends ran toward the center of the ring and huddled together again . . . but this time there was no escape.

CHAPTER 15
DARK LANDINGS

The Dragon of Death dived toward the friends until she was so close that Billy could smell her toxic breath. He thought he might be sick.

"This will be the last game you ever play," she said. "I will enjoy knowing you are suffering in the farms."

The walls around the four friends crashed down, and Billy was plunged into crackling darkness. He felt as if he were being sucked into the icy depths of the ocean as he fell deeper and deeper, and his lungs burned as if he'd swallowed hot coals. He tried to breathe, but nothing came. He tried to scream for his friends, but the more he tried, the more his lungs felt like they might explode. He swung his arms like windmills, trying desperately to find Ling-Fei, Charlotte, or Dylan, or anything to stop him from being plunged further into the dark.

Billy felt as if he were being poured from a glass.

He tumbled down and landed facedown on what felt like thick, gloppy mud. Billy gasped and rolled over, trying to catch his breath.

"Hello?" Billy managed after a few moments. "Guys? Are you there?" Billy blinked a few times, trying to adjust his eyes. It was dark, wherever he was, even though it had been daylight when they had fallen into the ruby sea.

"I'm here," said Charlotte in a weak voice, coming from out of the dark next to Billy. "But I can't see anything."

"I'm here too," said Ling-Fei from Billy's other side. "This place feels terrible!"

"Are we dead?" asked Dylan. "I feel like I died."

Billy sat up, but it was difficult in the slick and sticky sludge that they had fallen into. As his vision returned, he realized with relief that his friends were planted right next to him. He also saw that the ground underneath him, and as far as he could see, coursed with electricity, and strange shadowy shapes loomed in the distance.

He felt a weird sensation all over, as if he were covered in leeches. Billy inspected his immediate surroundings more carefully and saw that electric roots were swarming all around him. They came closer and closer until he couldn't get away, darting into his skin like needles and slowly pulling him deeper and deeper into the ground.

"Get off me!" Billy yelled, trying to shake the roots off. But the more he moved, the faster the electric roots pulled him down.

Billy realized with sudden, shocking horror that the round shadowy shapes surrounding them were rows and rows and rows of heads, both dragon and human, poking out of the ground like cabbages.

"I'm being sucked in the ground by creepy electric roots!" cried Dylan. "Somebody think of a way to get us out of here!"

Billy's thoughts raced in his head. There had to be something they could do.

"Yuck! Get off!" yelled Charlotte, who was fighting so hard to get free that she was already waist deep in the ground. "These things are like gross alien veins!"

"That's it!" said Billy. "Veins! These must be connected to the life veins. Quick, grab my hands, we don't have much time."

With effort, his friends came close enough so they could grab hold of him. Billy closed his eyes and reached for the same prickling heat that overcame him when he traveled through the life veins each day to get to their train car. As he reached for it, the familiar weightlessness came over him.

"It's working," Billy cried. "Hold on tight!" And without another breath, the friends were flying through the familiar rainbow current of light that Billy associated with the life veins. But this time, Billy didn't know where they were going to end up.

With a burst of speed, they landed with a thump on a surprisingly soft and bouncy surface. They all stared at each other in shock. "We're alive," said Billy, a bit incredulously. Even though he didn't know where they were, they were together. And that was the important thing.

Dylan was the first to pop up on his feet. "*What was that?*" he said as he patted himself all over. "Are we okay? Where are we now?"

Charlotte sprang to her feet and muffled Dylan's mouth with her hand. "Quiet!" she hissed. "We don't know who or what else might in here with us."

Billy saw that they were in a huge cavern. The air was so sour it stung Billy's eyes and nostrils. Above them, a massive sphere of electricity hung like a disco ball from a vaulted ceiling, sending light dancing between mirror-covered walls. Billy stared up at it, frowning. "I think we flew out of that thing." He realized that it was a direct line from the farms to the Tower—a kind of miniature portal.

"This place looks like a party house," said Charlotte.

Dylan shuddered as he pushed his glasses back up on his nose. "Well, this isn't a party that I would ever want to go to." Then he hopped up and down. "Whatever we've landed on is so . . . bouncy. What's underneath us?"

Billy scratched his head. "I think it's some kind of giant . . . bed?"

"Oh, no," said Ling-Fei. "I've got a very bad feeling about this."

Billy felt a gust of wind and realized that what he thought was a wall was actually a giant purple velvet curtain. He caught a glimpse of the darkening blue sky behind it. It looked close to sunset, which didn't make sense because it had been dark at the farms. But perhaps farms were dark no matter what time of day it was. A shiver shot down his spine. He looked around the room for more clues and saw that a pile of winking jewels and gold medallions were stashed in the corner of the room.

"I do too. I think we're in a dragon's lair."

"We're not just in any dragon's lair. I think we're in the Dragon of Death's lair at the top of the Tower," added Ling-Fei, who was as pale as the moon. "There is a terrible dark magic in this place, and I can feel it getting stronger."

Another gust of wind rushed through the room, this time so strong the curtains flew wide open, and the Dragon of Death emerged, power crackling all around her.

CHAPTER 16
FRIEND OR FOE

Before Billy could react, he was knocked down from behind. Charlotte had tackled all three of them and brought them into the bed. She was holding her finger up to her lips. Billy held his breath. The Dragon of Death must not have seen them! If she had, she surely would have attacked them immediately.

The Dragon of Death landed with a swoop in the cavern. "Oh, what stupid creatures humans are," she said to herself. "They will do anything for a dragon bond. It is all so *exhausting*," she huffed as she slunk toward the bed.

Billy could feel Dylan shaking next to him. He silently wished Dylan would maintain his nerve. The last time they were under this kind of pressure, Dylan ran, and they were almost eaten by a magic tiger.

The Dragon of Death came to a halt. "That is a funny smell." She sniffed the air. "Is that . . ." The dragon took a step toward the children.

Suddenly, three impossibly bright lighting strikes flashed like cracks in the dark blue sky. The Dragon of Death turned around. "Oh, what does Death's Shadow want now?" With a final long sniff, a suspicious glance around her lair, and an admiring gaze at her own reflection in a massive mirror, she flung her wings behind her and took off into the sky.

Billy exhaled. "That was close."

"Close? CLOSE? We are way past close!" Dylan exclaimed. "We fell into the farms and then you zoomed us STRAIGHT INTO THE DRAGON OF DEATH'S LAIR and she NEARLY FOUND US. And that means she nearly killed us! This is not close. This is closer than close. This is so close to being dead that we're probably getting rigor mortis as I speak."

"We're not dead yet," Charlotte quipped.

"But we should be! We should be dead a hundred times over."

"Dylan, we're okay," said Ling-Fei. "That's what matters."

"We aren't dead, but we're definitely NOT okay." Dylan began to pace in the lair. "I don't want to be here anymore. I don't want to fight any more dragons. I don't want to fight anyone! I want to be home in my own bed in Ireland. I want to be eating my granny's homemade brown bread. I never want to leave the house ever again." Dylan finished his speech in a shout, his chest heaving. The tiny gold pig, who had been safely in Billy's pocket, squeaked as if in agreement and flew to Dylan's shoulder.

"Will you keep it down!" hissed Charlotte. "You're never going to get to do those things if your shouting gets us caught!"

A slow clap echoed in the lair. Everyone reacted immediately.

Dylan let out a small squeak. Charlotte spun on her heel, fists clenched and ready. Ling-Fei tilted her head, listening intently. Billy took a step toward the sound, gazing into the dimly lit room, trying to see who was there.

JJ stepped forward from a hidden door in the wall. "For someone so intent on not dying, you're being very loud."

Ling-Fei gasped. "You! What are you doing here?"

JJ smirked. "I live here with my YeYe. Well, technically in the next room over."

"How awful," said Charlotte with a shudder. "I can't imagine anything worse."

"It isn't too bad. You get used to the screams and the smoke and these glasses help with the glare from all the energy sparking everywhere."

"So, you've really gone full evil," said Billy, shaking his head.

JJ tilted his head in Billy's direction and his brows rose up from behind his dark glasses. "Not as evil as your dragon, by the looks of it."

"You take that back!" Rage pulsed through Billy's body.

JJ snorted. "I think me insulting your dragon is the least of your problems." Then he slowly took off his black sunglasses. Billy gasped. Behind the sunglasses, JJ's eyes were so bloodshot they were practically glowing red. "I don't get much sleep. And with power radiating every-where all the time, it's hard to see. Even with the glasses."

"And here I thought you were just making a dramatic fashion statement," said Charlotte.

"I wouldn't be so rude if I were you," said JJ. "All I need to do is pull this bell . . ." He nodded to a purple cord dangling in the corner of the lair. ". . . And nox-guards will fly in here immediately."

"We can take you down before you reach it," said Charlotte through gritted teeth. "We're stronger than you. We always have been."

"Yes, yes, stronger together. Blah blah blah. I know, I remember. But here's the thing. Here? In this future, I've got a dragon on my side. And you don't."

"So, what are you going to do?" said Billy. "Why haven't you yelled for Old Gold or for Da Huo?"

JJ sighed and looked down at his hands. "This isn't the future I want, you know. I wanted to be with my grandfa-ther. But . . . not like this." Then he looked back up at Billy and the others. "What are you four even doing here? Why didn't you stay hidden? She would never have found you if you had stayed underground with the rest of the humans."

"Humans here can't be trusted," said Dylan, speaking for the first time since JJ had appeared. "Nox-hands will turn you in for a chance to get on the good side of a dragon."

"I can't blame them," said JJ. "It's a hard life for humans without any kind of protection from dragons. Da Huo is second in power only to the Dragon of Death herself and I'm still scared of her." JJ paused and seemed to think for a moment before looking at Billy. "Actually, your old dragon might be more powerful than Da Huo. Maybe even than the Dragon of Death. It's like there are two of her now. And the scariest thing is that she never shows any emotion. Death's Shadow will take humans and dragons to the farm or drain them of some of their life force without showing a hint of feeling." He looked thoughtful. "I wonder what she would do if she saw you."

"Nothing," Billy choked out. "She doesn't recognize us anymore. She stared right at me when she came for a street thief, and we were caught in the middle of it."

"That must have hurt," said JJ, shaking his head. "If Da Huo didn't know me, I don't know what I'd do. It would be like losing a part of myself."

"Well, you still have your big evil dragon buddy, don't you?" snapped Charlotte. "So, aren't you the lucky one?"

"Must be just as bad for you three, knowing your dragons are so close but that you can't get to them," said JJ.

Dylan frowned. "What do you mean by 'so close'? They're down in the dungeon and we're at the top of this giant Tower."

"Yeah, and where do you think that hole leads to?" JJ pointed to a huge hole underneath the massive bed that was covered in jewels and gold. "The Dragon of Death wants direct access to the dungeon. All the prisoners there are her personal life force source."

139

"Then that's where we're going," said Billy.

"You really must have a death wish," said JJ, shaking his head. "Anyone who goes into the dungeon doesn't come out. You die down there."

"Which is why we're rescuing our dragons," said Ling-Fei defiantly.

"You really think that you have a chance at saving your dragons? This is the Dragon of Death's destiny. She's made it impossible for anyone to ever escape the dungeon."

"Well, if we stay here, we're dead anyway," said Billy.

"If you go to the dungeon, you're as good as dead. Worse than dead," JJ replied.

"Can't you help us?" said Ling-Fei. "I know we haven't gotten along for a long time, but we used to play when we were kids. Remember?"

JJ looked away. "Even if I wanted to help you, I don't know how I'd get you out of here. The only way off this floor is by flight. And I don't have any nox-rings, or the power to fly on a magic cloud like my YeYe."

"That isn't the only way out," said Dylan. He pointed at the hole. "You said that goes to the dungeon, right?"

"I wouldn't call that a 'way out.' More like jumping from the frying pan into the fire."

"The dungeon is exactly where we want to go," said Billy.

"Then be my guest. I'll even do you a favor and not tell anyone that I saw you."

"How can we trust you? What if that hole leads somewhere else?" said Charlotte, scowling at JJ.

"Well, there is literally nowhere worse than the dungeon. So I don't know why you'd think I'd lie to you about where it is. And . . ." JJ trailed off. "You all kept me alive in the Dragon Realm when you didn't have to. You could have left me for dead and you didn't. There's nothing I can do for you here. I have no real power—only my YeYe has any sway over the Dragon of Death, and he certainly isn't going to want to help you. If you want to go to the dungeon, be my guest. Staying quiet about seeing you is the least I can do."

"You'd do it for Da Huo," said Billy quietly. "I know you would."

JJ nodded. "You're right, I would." Then he looked over his shoulder. "If you're going, you have to do it now. It's almost time for the Dragon of Death to take her nap."

"The Dragon of Death *naps*?" exclaimed Dylan.

"Of course she does. But she sleeps with her eyes open, so don't get any ideas. And then after her nap she'll want to go to the dungeon for her daily dose of life force, so you'll either see her there or up here."

"I'd rather face her with our dragons," said Charlotte.

"That's the other thing," said JJ slowly. "I don't know how much of your dragons will be left. They've been down there a long time. Nobody lasts long in the dungeon."

"No," said Billy, shaking his head adamantly. "I know for a fact that they're down there, and they're fighting to stay alive. To get back to us." He gave JJ a long look. "It isn't too late, you know. You can join us."

JJ sighed. "I hate to admit it, but you're all a lot braver than me. I'm not going to walk into the open jaws of death. Even if I wanted to defy both my YeYe and Da Huo, I don't think I could do it. I'm sorry." He looked down and away from Billy and the others, as if he were ashamed of himself.

"I didn't want you to come with us anyway," said Charlotte, tossing her hair.

"You can be good, you know," said Ling-Fei.

JJ scoffed. "I don't think I know how to anymore."

"You aren't calling the nox-guards on us," said Billy. "That's a start."

A small smile crossed JJ's face. "Consider it my good deed of the century. Now get out of here. I can't protect you if anyone comes in here. Especially not against the Dragon of Death."

"Well, thanks, I guess," said Billy. "Maybe we'll see you later."

"I doubt it. But good luck down there."

Billy and his friends stood at the edge of the hole under the massive bed. "We'll die if we jump," said Charlotte, staring into it. "It's too far down."

"I have an idea!" said Ling-Fei. She ran over to the huge windows that were hung with giant curtains. "We'll use the curtain as a parachute! Help me pull this down."

"That still sounds an awful lot like jumping," said Dylan.

"Ling-Fei is right," said Billy. "It's the only way."

They yanked down the enormous curtain as fast as they could.

"I sure am grateful that the Dragon of Death has taken such an interest in interior design," said Dylan.

"She's got terrible taste, though," said Charlotte, wrinkling her nose. "If this was my evil lair, I'd go in a totally different direction."

"You can design your lair another time," said Billy. "We've got to go." They hurried to the gaping hole beneath the Dragon of Death's giant bed.

"Do you really think this is going to work?" said Dylan.

"No idea," said Billy. "But it's our best shot."

"If I had a euro for every time you've said that," muttered Dylan.

"What's the worst that can happen? We die fighting to save our dragons? Save the world? We've always known those were the risks," said Charlotte fiercely. "I'm scared, but we have to try."

"We have to have faith—in ourselves and in our dragons," said Ling Fei.

"I'd take a working parachute over faith," said Dylan as they each grabbed a corner of the curtain. "But I guess this will have to do."

"No turning back now," said Billy. "Hold on tight! One, two . . . three!"

And the four friends leaped into the pulsing darkness.

CHAPTER 17
THE DUNGEON

Down, down, down they went, the four of them each holding one of the curtain's corners. With relief, Billy had felt air fill the fabric like a parachute and they floated slowly down the hole.

"I knew this would work!" said Ling-Fei, smiling.

"Yes, but, according to JJ, we're falling to our certain doom," said Dylan.

Billy saw a flash of gold shoot from his tunic pocket toward Dylan. The tiny gold pig was still with him. In all the excitement, Billy had almost forgotten that she was still there. Thank goodness she had stayed hidden when the Dragon of Death had flown into her lair. Goldie flapped in a zigzag path over to Dylan's face and licked him on the nose, as if to tell him everything was going to be okay.

"Oh, hey, little piggy," said Dylan, who was also smiling now.

"Hey! Focus," said Charlotte. "Do you feel that? In your chest? I think I can feel Tank!"

"You're right! I can feel Xing too!" said Ling-Fei who was beaming now.

"And Buttons!" said Dylan.

"We must be close," said Billy. "I can see something flickering below us! Get ready to land."

A moment later, the children hit the ground, the velvet curtain falling onto them.

"What new dark magic has sent the voices of our children to torment us," said a weak voice.

"XING!" cried Ling-Fei, throwing the curtain off her. "It's me! It's us! We're all here! We've come to . . ." Her voice trailed off and when Billy looked over to the dragons he knew why.

Their dragons were dying. They were slumped against the floor and their once bright gold eyes were dull and unfocused. The chains that were strapping them down clearly weren't making much difference as they looked as if they could barely move. Billy felt a sharp pang in his heart, and he knew it would be worse for the others, that each of them would be feeling their dragon's pain down their heart bonds.

"Oh, no," breathed Ling-Fei. "Are we too late? Xing! It's us!"

Xing managed a weak smile. "Can it be?" she huffed.

"Don't be foolish," said Tank, who still had his eyes closed. "There is no way our children could make it down here."

Charlotte rushed over to Tank and threw her arms around him. "It's us! It really is us!"

Tank's eyes brightened and he lifted his head. "Charlotte?"

"Yes, Tank. It's me! We've come to save you!"

Dylan bounded over to Buttons who was barely conscious. "Can you hear me, Buttons?" Dylan said. He ran a gentle hand over Buttons's snout.

Buttons shifted his head and opened an eye. "Human?"

"Yes, a human. And not just any human, Buttons. It's me, Dylan." Dylan threw his arms around Buttons's neck.

Buttons inhaled deeply as if taking in Dylan's scent. He breathed out a puff of gold smoke and when he opened his eyes, they were a shimmering gold.

Billy watched as Ling-Fei, Charlotte, and Dylan gently lifted their dragons from the depths of despair that had consumed them. He could see life slowly starting to come back to them through the strength of their heart bonds. As much as it pained Billy not to be with Spark, seeing his friends back with their dragons mended a small piece of his heart that had been broken when Spark did the unthinkable back in the Dragon Realm.

"Incredible," said Xing. "It really is you."

Ling-Fei was lying on top of Xing's head, snuggling her between her antlers. "Nothing was going to stop us."

"Well, nothing except for the thousands of evil dragons who are surrounding this place," said Dylan.

"Let us have our moment, you grumpy opossum," said Charlotte.

Dylan looked over at Charlotte and shrugged his shoulders. "I'd just like it to be known that there were many, many, *many* things that could have stopped us. And, yet, here we are."

"How did you manage it?" said Buttons.

"It's a long story," said Billy. "We don't have time to tell you everything. But we've made friends—friends we can trust."

"Interesting," said Xing. "I will admit that after what happened with Spark, I find it hard to trust anyone."

Billy hung his head, feeling as if he himself were responsible for Spark's betrayal.

"That isn't an attack on you, Billy," said Buttons gently. "We know you've been cut the deepest by Spark's actions. I'm sorry."

"If she fooled me, she could have fooled anyone," said Xing grimly.

"I wish . . . I wish I could have stopped her," Billy said, his voice quiet in the dim dungeon.

"You did everything you could," said Tank. "And there is no point focusing on how things could have been different. We need to focus on the now." His voice was rough, but his eyes were kind. "Dylan, you are right to say there is danger all around us. You should not have come here. There is no way out of this dungeon. We are trapped by the Flames of Death. Nothing can pass through them."

"Oh, how could I forget the death flames," muttered Dylan under his breath as he eyed the flickering purple flames that surrounded them from the floor to the ceiling.

"That must be what Thunder was talking about," mused Charlotte.

"But we have another way out!" said Billy, pointing to the hole they fell through. "We just need to break you out of these chains, and we can fly out of here. The Dragon of Death isn't in her lair, so if we're quick we can escape!"

"Billy is right," said Xing. "The Dragon of Death has grown arrogant. Now that we have reunited and been strengthened by our heart bonds, we might be able to break free."

"I was too weak to heal myself or the others before you came," said Buttons. "But now that you're here, Dylan, I think I may be able to."

Buttons closed his eyes and began to hum the familiar tune that Billy now knew so well. Clouds of gold flowed from Buttons and swirled around the group. A warming sense of peace and calm permeated Billy's body. He felt as light as air and his muscles relaxed as Buttons's magic healed him. When Buttons stopped humming, Billy opened his eyes. He felt fully rested.

"Well done, Buttons," said Tank. "Oh, how I have missed your healing power." Then he jerked his head back, tugging against the chains. "Stand back."

With a mighty roar, Tank burst out of his chains, his wings jolting open as bits of broken metal flew across the dungeon. "It feels incredible to be free and to be able to *stretch*."

The pig, who had been resting in Billy's tunic, emerged, startled by Tank's roar. It flew around the dungeon, zipping between the dragons and the children.

"I see that pig is still with you," said Xing flatly.

"I think she likes us," said Billy. "She follows us everywhere."

Tank plodded over to Xing and Buttons and bit down on the chains that held them down. With a quick flick of his neck, he snapped the chains off. "No time to waste," he said. "Let us fly before the Dragon of Death finds us."

"To freedom!" cried Buttons.

"Oh, it's so good to be reunited," said Ling-Fei. The group moved to the hole that Billy and the others had descended from and gazed up at it.

"Strange," said Xing, eyeing it. "It cannot be that easy." She tentatively flicked the end of her tail upward. As she did, crackling purple electricity shot out from the sides of the hole, sealing them in.

"How strange," said Ling-Fei.

"The Dragon of Death has enchanted this entrance," said Xing. "Only she can pass. You children are lucky that the Dragon of Death did not think to shield it from humans, or you would not have made it through."

"So what are we going to do now?" asked Dylan forlornly.

"There is no way out," Tank said with solemn finality.

"Are you sure we can't just fly *through* the death flames?" said Charlotte, eyeing the wall of flames around them.

"We can't risk it," said Buttons. "It's too dangerous. We've seen what happens when something goes through the flames."

"A nox-guard stumbled into them once, and in moments all that was left of him were charred bones. Even nox-wings cannot go through them," said Xing.

"You have come all this way only to die in the dark," said Tank gravely. "You should not have tried to rescue us."

"But we're together," said Billy. "There must be something we can do!"

The tiny gold pig squeaked as if in agreement, and then, before Billy realized what it was doing, it zoomed straight into the Flames of Death.

"No!" Billy yelled, lurching forward to try to catch it.

"Stop!" Xing cried, wrapping herself around Billy and pulling him back from the flames. "There is nothing you can do."

"The pig was just trying to help us," said Billy, fighting tears.

"I had also grown fond of the silly thing," said Xing. "But she has died a brave death."

"And it sounds like we'll be dead soon too," said Dylan glumly.

They all stared into the purple flames.

And then, through the flickering flames, a small round shape appeared. "Wait, what's *that*?" said Charlotte.

The tiny gold pig burst through the flames and oinked.

"You're alive!" Billy shouted. The pig squeaked indignantly and then quickly turned around and flew back through the flames, before zigzagging back again.

"Now it's just showing off," said Charlotte.

"How is that possible?" said Ling-Fei.

"*Oh, of course.* The pig is an immortal creature. I should have sensed it. It must have eaten a peach of immortality," said Xing.

"How rare," said Tank.

"Wait!" said Dylan. "The peaches of immortality are a *real* thing?"

"Of course they are," said Buttons. "Why would we have lied to you about that?"

"It just seemed kind of impossible," said Dylan.

"More impossible than dragons?" asked Billy.

Dylan nodded. "Good point. But do you know anything else that has eaten a peach of immortality?"

"The only other being we suspect to have eaten one is the Dragon of Death," said Xing. "Which is why we cannot kill her, even if we managed to overpower her. The only thing we can do is send her into a black hole."

"Ah, yes, the old 'Send them into a black hole' trick," said Dylan with a grin. "It sounds easy, especially from inside a dungeon, trapped behind a wall of death flames. Even if the pig can go through them, we still can't."

Billy found himself slowly stretching out his hand toward the flames, but the heat was so strong that he jerked back before his hand even got close. "Yep. Those flames will definitely burn us," he said. But there had to be some way. If the pig could go through them . . .

He suddenly remembered the pig eating the Life Bar wrapper—something it definitely shouldn't have been able to eat. But now he knew why it could eat anything . . . it was an immortal pig.

"If the pig can go through the flames, do you think she can . . . eat them?" said Billy slowly.

"Eat the flames?" repeated Charlotte.

"Exactly!" said Billy. "Like . . . slurp them up?"

The dragons glanced at each other and then at the pig. The pig oinked.

"I suppose she could try," said Buttons.

"Eating the flames won't hurt the pig, will it?" said Ling-Fei with a worried frown.

"I do not know the effect of enchanted flames on an immortal flying pig's stomach," said Xing. "I do not think anyone knows."

"Well, we're about to find out," said Charlotte. "Look! The pig must have understood us!"

The tiny gold pig flew to the base of the wall of flames and began to snarfle and snort, slurping up the flames.

"It's working!" cried Billy. And it was. The flames were getting lower and lower the more the pig devoured them. Soon, the entire wall of flames was gone and only the dungeon walls remained.

"Wow," said Billy. "Good job, Goldie! Come here." The pig flew over to Billy's outstretched hand, swaying a bit in the air.

"She looks especially full," said Dylan, sounding alarmed, "like a balloon that might pop."

The pig landed on Billy's hand, squeaked, and then burped out smoke.

"Whew," said Billy, wrinkling his nose. "That stinks."

The pig burped again, yawned, and closed her eyes. Billy tucked her into his pocket, and she began to snore softly.

"That solves that problem," said Charlotte. "What now? Do we try to fly out the way we came in?"

"Back through the Dragon of Death's lair?" Ling-Fei shook her head. "I don't think that's a good idea. She could come back at any moment and catch us."

"Well, I don't see any doors or windows," said Dylan.

"Dragons do not need doors," said Xing smugly. "Ling-Fei, get on my back."

"Dylan, you're with me," said Buttons.

Charlotte climbed up on Tank. Billy stood a moment, feeling like a spare part. He missed Spark—the Spark he'd known and loved with sharp fierceness.

"Billy, what are you waiting for?" said Charlotte. "Get up here with me."

"We cannot do this without you, Billy," said Tank. Feeling buoyed, Billy clambered up onto Tank, sitting behind Charlotte.

"Now, hold on!" Tank shouted. With a mighty roar, he charged headfirst toward the wall of the Tower dungeon.

CHAPTER 18
THE STRENGTH OF A SHADOW

Bones and bolts of electricity flew through the air. The Tower walls were fortified with electricity and power—the same power that ran through the life veins. Despite Tank's efforts, the wall still stood. But now, wave after wave of power radiated from the battered dungeon walls, blanketing the entire room with pulsing energy. Billy felt it settle into his bones, and a moment later, he realized his hands were glowing.

"Whoa," said Charlotte. Billy looked up and saw that her hair was standing on end, surrounding her like a halo. He glanced back over his shoulder. Behind his glasses, Dylan's eyes shone, almost like they were emitting their own light.

"Did you feel that?" yelled Ling-Fei, who was glowing with pulsing power.

Billy raised his hands and reared back as small blue

sparks jumped from his fingertips. Then he realized all the dragons were crackling with the same electric power he usually associated with Spark.

"Xing, Buttons—I need your help!" Tank bellowed.

"We're with you, Tank!" called Xing.

"Ready!" shouted Buttons.

"Children, focus your new electric powers on the wall. Those waves of power have made you stronger!" said Tank.

Billy felt his newfound power rising up inside him, as if it were answering Tank's call. He focused as hard as he could and held his hands out in front of him. An electric blue blast of power emerged from his fingertips and shot straight at the wall. Dylan and Ling-Fei's power streams rushed by him. Dylan's was bright green and Ling-Fei's a shining silver. Ahead of him, Charlotte's red power stream hit the wall so that they were all directing their energy at the same spot. The wall of the dungeon buckled in the middle, right where their combined powers were blasting.

"GO!" bellowed Tank, and with his head tucked down, he barreled straight into the weakened wall. Xing and Buttons rammed into it at the same time, and they burst through into the night.

"Oh, it feels delicious to fly again!" Xing cried out, uncurling her body to its full length. She undulated in the sky like a wave in a shining sea.

"Air! Fresh air!" shouted Buttons, mouth wide to breathe in as much of the night as he could.

"Stay focused!" roared Tank. "It looks like we have company."

It was true. Despite there now being a massive hole at the base where the dungeon was, the Tower still stood. Nox-wings were emerging from every window and Billy saw the hot pink nox-wing emerge from the wreckage of the dungeon and blast toward them, its spiked tail spinning behind it like a helicopter blade.

"Fly higher!" cried Xing.

But they were already too close to the ground and there were nox-wings coming at them from every direction.

Suddenly, there was an almighty screech from above that shook the air. "STOP THEM!" It was the Dragon of Death, diving from the highest point of the Tower. Her wings were flung back as she shot through the sky like a bullet headed straight toward them.

In that moment, Billy's heart plummeted. There was nowhere for them to go. They were surrounded by nox-wings. It had all been for nothing. They'd escaped only to be caught again. It didn't matter that they now had electric powers. They were no match for the Dragon of Death. She would win again.

Then, a familiar shadow swooped through the air. Faster even than the Dragon of Death herself. It easily caught up to her, and when it did, it opened its wings to their full enormous length and wrapped itself around the Dragon of Death. It was Spark's shadow.

The Dragon of Death screamed and clawed at the shadow. But it was no use, her talons went through it like smoke and the shadow simply rearranged itself around her. The nox-wings all froze, staring at the Dragon of Death wrestling with the shadow. The shadow that everyone thought answered to her. From further above, Spark watched her shadow take on a life of its own. And then she looked down and locked eyes with Billy.

A surge of warmth ran through him, and then, louder than he had remembered, Spark's voice echoed in his mind. *Quick, Billy, fly away! And fly fast. I cannot hold her back for long.*

"Spark!" Billy cried out loud, unable to contain himself. But she had already turned her attention back to the Dragon of Death.

And then, a deeper darkness than Billy had ever known fell over them. It was so dark, Billy thought he could feel it. It felt like velvet. It blanketed all the sound too. Except one voice, very close to Billy's ear.

"I thought you might need my help!"

It was Midnight!

When the world flickered back on, Billy saw that he and his friends, and their dragons, were in some sort of giant black bubble. The tiny gold pig squeaked in excitement from inside Billy's tunic pocket.

"They can't see you," Midnight carried on. "I've put you in my Midnight blanket!"

"Who is this?" growled Xing.

"I could swallow you whole!" thundered Tank.

"Wait!" cried Ling-Fei. "She's a friend."

Midnight glared at Tank. "I'd like to see you try to swallow me!" Her horns began to take on a warning red glow.

"Nobody is swallowing anybody," Billy said quickly. "This is Midnight. And I think she's just made us all invisible!"

It was true. Below them, the nox-wings flew frantically in circles, trying to see where they had disappeared. And to Billy's horror, he saw that the Dragon of Death had burst through Spark's shadow. The shadow looked limp as it re-joined Spark's body. Even from a distance, Billy could tell that Spark was weak from sending her shadow after the Dragon of Death. Her wings were flapping slowly, and her head drooped. Why wasn't she flying away?

"A traitor in my midst!" screamed the Dragon of Death. "A traitor who would dare try to stop me!"

The nox-wings took up the chant. "TRAITOR! TRAITOR! TRAITOR!"

"Traitors do not deserve to call themselves dragons! They do not deserve to fly!" The Dragon of Death soared upwards toward Spark who was hovering listlessly.

"HOLD HER!" the Dragon of Death shrieked.

Two nox-wings immediately grabbed hold of Spark's wings with their jaws, pulling them taut.

"Come on, Spark! Fight back!" Billy murmured. But Spark let herself go limp between the nox-wings, her head down.

Spark! Billy thought down their bond. *Spark! Do something!*

At this, Spark lifted her head the smallest amount. Billy could have sworn he saw a sad smile on her face.

Billy. My Billy.

Spark was still smiling when the Dragon of Death ripped off her wings.

CHAPTER 19
A DRAGON WITHOUT WINGS

"SPARK!"

Billy felt Spark's pain echo in his bones. He had to get to her! He had to save her! She was plummeting toward the ground in a blue and silver blur, gaining speed as she fell. The Dragon of Death's cruel laughter reverberated through the air as Spark silently dropped to the ground.

"Go! Go catch her! One of you!" Tears streamed down Billy's face as he kicked Tank in the side, trying to get the giant dragon to leave their protective Midnight bubble and save Spark. But Tank stayed where he was. Billy frantically looked to the other dragons. "Please!"

"I am sorry, Billy," said Xing. It was the gentlest Billy had ever heard her speak. "We will not reach her in time."

Billy looked at Buttons. "Buttons! You can heal Spark, can't you? We have to go to her!"

Buttons shook his head sadly and looked away. "I'm sorry, Billy," said Dylan.

Spark was still falling. Billy prepared to leap off Tank's back. He'd fall to the ground himself if that meant he could get to his dragon. He had electricity running through his veins, didn't he? Maybe he could use his new power to fly through the air. He had to do something. He'd risk anything to try to rescue her.

"Billy! You can't save her." Charlotte turned around from where she sat on Tank and gripped Billy's wrist. Billy squirmed and pulled his arm away, but it was no use. Charlotte was too strong.

"She's going to die! She's going to land and die and none of you are doing anything to stop it!"

"Spark is strong. She will survive the fall," said Xing.

"Don't watch," Ling-Fei said. But Billy couldn't tear his eyes away. He burned with fury at his friends and their dragons. Then Spark hit the ground like a meteor. She landed in the center of the Dragon Court in an explosion of dust and ruby rubble.

"No!" Billy cried, still fighting against Charlotte's iron grip.

"You'll feel if she hasn't survived the fall," said Buttons quietly. "Can you feel her through your bond?"

Billy closed his eyes and listened in his heart for Spark. *Spark? Are you okay?*

There was a long silence . . . followed by Spark's voice. It was so quiet, Billy almost didn't hear it. *Billy.*

You are still too close. The Dragon of Death will sense you. Leave. Please. Do not let my wings be gone for nothing.

I can't leave you, Spark.

You must. I will be fine. Do not worry. Protect yourself and the others. I believe in you, Billy.

Billy drew a deep shuddering breath. "She survived the fall," he said numbly. "But she wants us to leave while we still can."

"She is right," said Tank. He looked to Midnight. "Young dragon, can we travel in this . . . darkness you have covered us in?"

Midnight blinked. "I don't know. I've never tried to blanket so many at once."

"You can do it, Midnight," said Ling-Fei encouragingly.

"I can help you," said Xing. "I do not have your unique power, but if you ride on my back with my human, you can focus on the dark."

As Midnight flew past Billy, he noticed that she was trembling. "I'm sorry you had to see that, Midnight," he said.

Midnight held her head high. "I might be young, but so are you. You're out here fighting for what's right. I want to do that too."

"Well, I'm certainly glad you came to find us," said Charlotte. "Otherwise, we'd be toast."

"I had a feeling you might need me," said Midnight. "And I thought I'd find you near the Tower."

"You saved us, young dragon," said Buttons. "You and Spark's shadow. Tonight, we were protected by the dark. We owe you a great debt."

"Be sure to tell my father that when we return to our lair," said Midnight, settling herself on Xing's back behind Ling-Fei. "He's going to be very unhappy with me."

Midnight kept them under the cover of darkness until they reached the marble orb. As they approached it, Xing inhaled deeply. "There is strong magic here," she said, giving Midnight a quizzical look. "Who did you say your father is?"

"You'll see," said Midnight. They landed on top of the marble orb as a group. The building recognized Midnight right away and a door slid open underneath them.

Once they were safely inside, and the roof had closed again, Midnight dropped her shield and collapsed with exhaustion on the ground. The tiny gold pig immediately flew out of Billy's pocket and went over to Midnight, buzzing around her and squeaking in concern.

"WHAT HAVE YOU DONE TO MY DAUGHTER?"

Thunder's voice was so loud that the entire marble orb shook with the force of it. Midnight sat up and yawned. "Father, I'm fine. I saved the day! But now I want to go to bed."

Thunder rushed over to Midnight, his long moustache and beard brushing on the ground. He stood in front of her in a protective stance, glaring at Tank, Xing, and Buttons.

Billy stepped forward. He still felt numb from seeing what had happened to Spark, but he wanted Thunder to know what Midnight had done for them. "Midnight saved our lives using her blanket of darkness," he said hoarsely. "She was so brave."

"I assume these are the prisoner dragons?" said Thunder, still eyeing them up and down.

"Prisoners no longer, thanks to your young dragon," said Tank. "And you must be the new trustworthy friend our humans have told us about."

Buttons cleared his throat. "If you'll allow me, I can heal your young dragon. She's exhausted from exerting so much power."

"She is very strong for a dragon so young," chimed in Xing.

"Very well," said Thunder.

Buttons began to hum his healing song, and even though Billy didn't want to feel better—he wanted to feel wretched, he deserved to feel wretched—in a few moments, not only did his body not ache, he also felt a weight lift off his shoulders.

"How did you escape the dungeon?" asked Thunder.

"Can you tell him?" said Billy, looking to his friends. "I . . . I don't think I can say it all out loud."

And so Ling-Fei, Charlotte, and Dylan took turns explaining to Thunder how they had snuck into the Tower, made it through the Flames of Death, and escaped the dungeon. When they got to the part about Midnight covering them in her darkness blanket, Thunder shook his head in awe.

"You have shamed me today. I should have been there too. We have hidden from the Dragon of Death for long enough. But no longer." Then his eyes landed on

Billy. "My Midnight was not the only brave one tonight. I never thought Death's Shadow would turn on the Dragon of Death. Your bond must go very deep indeed."

Billy nodded. "I have to go see her," he said. "I have to make sure she's okay. I've waited too long already."

"You aren't going back there," said Charlotte at the same time as Dylan emphatically shook his head. "Spark wanted you to be safe. You can't walk right back into the jaws of the Dragon of Death just for another glimpse of Spark."

"But she's hurt and weak!" said Billy. "I can feel it!" Then he let out a grim laugh. "At least now I know she won't be heart-bonded with any of the winners of the tournament. We're still bonded. I felt her pain when her wings were clipped."

Xing snapped her head up. "Wait, what do you mean? The winners of the tournament cannot be bonded with any dragons. That is not how the bond works."

"The Dragon of Death claimed she had come up with a new superior bond. She said the tournament would be how she worked out which humans she could bond with her top dragons," said Ling-Fei.

"No, no, no. That's impossible," said Buttons. Then he gasped. "Of course. She's evil, but she's clever. She's using it as a trap—a way to find the humans who are the strongest. She'll take them as her own personal life force."

"There is no way to fake a heart bond," said Tank. "The Dragon of Death knows that the more dragons who have heart bonds, the more powerful they will be. She will not want anyone else growing in strength and power."

"I should have known!" said Thunder. "My judgment has been clouded by my fear for my family."

"The grand finale is tomorrow," said Ling-Fei. "We have to stop it from happening. All those humans are going to be wiped out!"

"How can we stop her?" said Dylan, taking off his glasses and wiping them on his shirt. "She's so powerful."

"But so are you four," said Thunder. "You have grown in power since I saw you last. Whatever happened in the dungeon has made you stronger than you realize."

"Thunder's right," said Billy. "We're strong and we can easily blend in. We've got to go back. We're not trying to take down the Dragon of Death; we're just trying to save the humans who've entered the tournament."

"I will not allow it," said Tank. "We need to come up with a different plan."

"We need to save those humans!" Billy's voice broke. "It's our fault that Dragon City even exists. We've done enough waiting and hiding."

"It is too dangerous," said Xing. "We will be spotted right away."

Billy shook his head. "No, *you'll* be spotted right away. But we won't be." He gestured at Ling-Fei, Charlotte, and Dylan. "We'll use our disguises and Dylan's charm. We'll almost be invisible."

"And they will not be alone," said Thunder. "I have hidden in fear long enough. I will go with them. After all, it is expected that I would be at the tournament. Everyone in Dragon City believes I am a nox-wing. I should be there."

"Then it's settled," said Billy. "We're going to the tournament with Thunder. When we get there, we'll do whatever it takes to save the humans." *And Spark*, he thought. But he didn't say that out loud.

"We can't sit inside and do nothing when we have a chance to save so many people," said Charlotte. "I'd never forgive myself."

"Loyal, brave, strong, and true," said Buttons with admiration. "You still have the same hearts you did when you first opened the mountain and found us."

"And I still believe you have hearts strong enough to save us all," said Tank solemnly. "I do not like the idea of being separated, especially after we have only just been reunited, but I see no other choice. We are no good to you if we are captured again or, worst still, killed."

"There is something else I should tell you about," said Thunder. "Or perhaps I should say 'someone.'" He quickly explained about Lightning.

"We will watch over her while you are away tomorrow," said Xing. "Do not worry; no harm will come to her."

"It was wise of you to have kept her sleeping," said Buttons, "but the time may come when you'll need to wake her. Her strength could be what we need to defeat the Dragon of Death."

"I would if I could," said Thunder. "But she is in too deep of a sleep. Only another lightning dragon can wake her now. And I have never met another."

"I know of one," said Billy slowly. He looked at the human and dragon faces around him. "We need Spark. I know she betrayed us before, but she's clearly on our side now! I think she's been on our side this whole time, or at least tried to be. You saw what she did for us. With Spark on our side, and Lightning, surely we'll be able to defeat the Dragon of Death."

"I don't know if I can trust Spark again," said Charlotte. "I'm sorry, Billy. I saw what she did for us, and I'm grateful, but what if she turns nox on us again? You saw her take life force, and it didn't even affect her."

"She won't," said Billy, tears of frustration building in his eyes. "We need her!"

"It is a risk to trust her again," said Xing. "Charlotte is right."

"It's worth it if it means we can wake Lightning and destroy the Dragon of Death," said Ling-Fei.

"Okay, let's assume we're able to do all of that. Then what?" said Dylan. "Do we just live in this time forever?"

"We can figure out how to get home after we've destroyed the Dragon of Death," said Billy firmly. "One thing at a time."

"Grand," said Dylan dryly. "I'll just add 'Destroy the Dragon of Death' to my to-do list, shall I?"

And despite everything, Billy laughed. He felt buoyed with hope. They were going to save Spark. They were going to save everyone.

CHAPTER 20
THE GRAND FINALE

The grand finale of the tournament was taking place back where it had all begun—in the Dragon Court. But today there was no ruby floor. Instead, a long chute extended from the Tower itself. The hole in the base where the dungeon was located had already been repaired, and there was no sign that it had ever been wrecked at all.

Billy, Ling-Fei, Charlotte, and Dylan blended in with the crowd of spectators, and Thunder joined a group of nox-wings at the back. In the center of the court, the Dragon of Death stood on a raised platform. Next to her was Old Gold and beneath her, lying in a crumpled heap on the ground, was Spark.

"See how the mighty have fallen! Even Death's Shadow is at my mercy!" roared the Dragon of Death. She lashed out at Spark with a claw, and Billy cringed as he saw her gold blood pour out of her.

"No longer will the winner of the tournament be bonded with this poor excuse of a dragon. No, the winners will be able to choose among my hand-picked dragons!" The Dragon of Death flung her wings open and dozens and dozens of nox-wings emerged from the Tower and lined up. Billy saw one lick his lips hungrily.

The nox-wings knew. They knew that the humans who finished the tournament were to be used for life force. Nobody was being bonded. It wasn't possible. The whole thing was a setup.

"To be a bonded human is the highest a human can hope to climb in Dragon City," the Dragon of Death went on. "To have the protection of a dragon is much more powerful than being a nox-hand. It is what all humans should dream about."

The crowd cheered in excitement and anticipation as drums began to pound. The first humans began to tumble out of the long chute coming down from the Tower.

"We have to do something," hissed Charlotte.

"What do you suggest?" said Dylan.

Suddenly, there was a jostling behind them as a woman bumped into them. It was the nox-hand with the silver hair, the one who had let them go on their first night in Dragon City. She frowned at them. "Why do you kids look familiar?" She grabbed Billy's hand. "And how come you have lightning power?"

She yanked Billy out in front of the crowd. "This kid is radiating some kind of power!" she shouted. "I think he's got a nox-ring on him!"

Billy began to feel light-headed. He didn't know what to do. Any moment now the Dragon of Death would turn and see him. But she was still parading around on her platform, telling the crowd how wonderful and benevolent she was. She hadn't even noticed the disturbance. But other nox-wings had.

"I know that stinking human!" roared a dragon from the base of the Tower. It was the copper-colored nox-wing. In a fury, it took off in flight, heading straight for Billy. "I won him in the Dragon Death Drop! He belongs to me!"

The silver-haired nox-hand tightened her grip on Billy. "Well, you shouldn't have lost him!" she said. "Fair is fair. I'm a nox-hand, and I caught him. I'll sell him back to you."

"Do your business later!" shouted a voice from the crowd. "We are here for the tournament!"

But more and more people were turning toward Billy, the nox-hand, and the copper-colored nox-wing. Maybe this was the perfect distraction so his friends could help the humans who had survived the tournament and would be zooming out of the chute any moment, waiting for a prize that would never come. Billy looked back at Ling-Fei, Charlotte, and Dylan. "Now," he mouthed.

Charlotte nodded and grabbed Ling-Fei's and Dylan's hand. Billy watched as they snuck to the end of the chute, their hands sparking with power.

Billy knew he had to do something big to keep the crowd and the nox-wings' attention on him. He turned toward the copper-colored nox-wing, lifted his hands, and let his own power fly toward the dragon.

The nox-wing wasn't expecting that and flew backwards, barreling into the crowd of dragons. "Watch where you're going!" snapped one of them.

"That kid blasted me!" exclaimed the copper-colored nox-wing. The other dragons guffawed. "A human cannot do that. The only human with that kind of power is Old Gold, and he does what the Dragon of Death tells him to."

An ear-splitting roar broke through the commotion. Thunder dashed through the crowd. "You are mistaken," he said, looking down at the copper-colored nox-wing. "This human belongs to me."

The silver-haired nox-hand dropped Billy's arm. "You're not worth the trouble," she muttered and slipped into the crowd.

"Billy, what are you doing? You are going to get yourself killed!" whispered Thunder in a surprisingly quiet voice.

"We've got a plan," Billy whispered back.

"I want the human!" said the copper-colored nox-wing. "And I will fight you for it! It is a small human, but I won it! And you are old and will be easy to defeat."

Then, before Billy realized what was happening, the copper-colored nox-wing whipped its heavy tail against the side of Thunder's head. Thunder blinked as if in a daze, before slumping to the ground.

"Thunder!" Billy rushed to make sure he was still

breathing and heaved a sigh of relief when he saw the old dragon's chest moving up and down.

"Now you are coming with me," said the copper-colored nox-wing, advancing on Billy.

But then Billy heard a screech that chilled his blood and stopped the nox-wing in its tracks.

"YOU FILTHY FOUL HUMAN CHILDREN! STOP RUINING MY DESTINY!" It was the Dragon of Death. She had spotted Ling-Fei, Charlotte, and Dylan.

Summoning all his courage, and with the hope that he still had enough agility power, Billy sent out a blast of energy strong enough to propel him over the heads of the nox-wings and even the Dragon of Death herself. He landed on his feet in front of his friends, and the crowd gasped in awe.

Billy looked straight into the Dragon of Death's eyes as she flew toward him, eyes burning with hatred and power, but he stood firm. He wasn't afraid anymore. He felt his new electric power thrumming in his veins, and he held up his hands. At least he could go down fighting, at least he could try to show the people of Dragon City that they could fight back.

Something blue and heavy knocked him backward.

It was Spark's tail. She had somehow gotten between him and the Dragon of Death. Even without her wings, she was magnificent and powerful. She flung her head toward the Dragon of Death, opened her mouth, and blasted her with a beam of pure power.

The Dragon of Death flew across the court, slamming into the base of the Tower. The crowd grew deathly silent.

"Stay behind me," Spark said to Billy, her voice breathy.

"I can help you!" Billy put his hand on Spark's leg. "We're stronger together, remember?"

But before Spark could reply, the Dragon of Death rose up into the air and dived at them. Spark reared up on her back legs and sent out a blast of power, right as the Dragon of Death shot out her own stream of dark magic. The two streams of power hit in the middle and then exploded, sending sparks of electricity everywhere.

Spark was flung back, but her power had hit its mark. Billy saw that the Dragon of Death had been struck. She screeched in agony and then she glared at everyone assembled.

"You will all heal me!" she shrieked. "I will take as much life force as I need!" Billy tensed, waiting for her to unleash the noxious gas that would drain them all of life force and, for a moment, it appeared as if she were trying to. She flapped her wings, once, then twice . . . but nothing happened. The crowd began to murmur, and Billy saw something like panic flash across the Dragon of Death's cruel features. She looked around frantically until her gaze landed on Da Huo. "Finish them!" she shouted. "The tournament will continue when I return!" With obvious effort, the Dragon of Death took off for the top of the Tower, Old Gold in her wake on his floating cloud.

The murmuring crowd exploded into noise as it became clear that nobody understood what had just happened. There were still tournament participants coming out of the chute and landing at the base of the Tower with dazed expressions on their faces.

But Billy wasn't watching any of that. All he cared about was Spark.

"Billy," she said softly, and then she collapsed in front of him.

CHAPTER 21
GOODBYE

Gold blood poured out of a gaping wound in Spark's chest. With each breath she took, she bled more and more. Her head lay on the ground, as if she didn't have the strength to lift it.

But worse than the blood was the ash. Her scales were starting to turn to ash, flaking right off her skin, and the sight of it pained Billy terribly. There was something so wrong about it. How could he finally be reunited with Spark, his Spark, just for her to turn to ash? What was happening?

The crowd behind them had gone very still, but Billy was barely aware of them. All he could focus on was Spark. He drew closer to her so he could see her face. Her eyes were closed, and her breathing was labored, and with each second, more and more of her scales turned to ash. With shaking hands, Billy gently stroked between her antlers.

"Spark! You're going to be okay. Buttons can heal you." He said it as if it were the truth, because it had to be. Spark *had* to be okay.

Spark's eyes fluttered open, and to his enormous relief, Billy saw that they were no longer black, but gold again. "Billy, it is so good to see your face again. I have been watching you, trying to protect you when I could. I led you to the train where I knew you would be safe, and in the Dragon Death Drop, it was me who summoned the wind."

"I knew you were never fully nox," said Billy, trying to swallow the lump in his throat. He had to be strong for Spark.

"I let my hunger for power, for greatness, overtake what is most important. Goodness. But you were never tainted. You stayed good. I am so proud to call you my human." Spark gazed around at the Dragon Court. "Look what I have done. All because of my addiction to power, to dark magic."

"I know you have a good heart, Spark. Everything is going to be better now. You're going to live and we're going to win!" Billy hated that his voice was shaking, but he kept going. "We can fix it."

"You do not need me to save the day, Billy. I have turned into the villain." Spark began to cough, and gold blood sprayed out of her mouth. "I am so glad that I can now tell you how sorry I am. For everything. I do not even deserve to say goodbye to you, but I am grateful that I am able to."

"We're not saying goodbye," Billy's voice cracked.

"Billy, you will do great things. This is the start of your adventure. Promise me you will stay loyal, strong, brave, and true."

"Always," said Billy, as tears began to fall from his eyes. Spark's breath was coming more slowly now, and her scales were continuing to flake off. She was growing smaller right before Billy's eyes.

"My brave Billy," Spark said, her voice raspy. "I believe in you."

And then she closed her eyes and her breathing stilled.

Now Billy knew why Buttons had said he'd be able to tell if Spark died. Suddenly his whole body was filled with a terrible ache, as if something were missing, as if his own heart had been ripped out. He gasped, putting his hands on his chest, almost as if he thought there might be a hole in his body.

"Billy!" Ling-Fei, Charlotte, and Dylan emerged from the crowd and pulled him back from Spark's prone body. He kicked and elbowed them, but they were stronger than him and all he could do was cry.

As Billy watched through tear-blurred vision, Spark turned entirely to ash, until all that was left of her was one shining blue scale. Ling-Fei darted out and gently plucked the scale from the mountain of ash, bringing it back to Billy. He took it and clutched it in his hand, trying to comprehend what had just happened.

Spark was gone.

Forever.

CHAPTER 22
THE CHASE

Billy couldn't stop shaking. He fell to his knees as a sob tore out of him like a wild thing. He clutched the blue scale—all that was left of Spark—and shook so hard that his teeth chattered. Nothing, not even having his life force drained, hurt as much as this did. It felt as if his whole body were made of lead, as if his brain were full of cotton. He almost expected to turn to ash himself. He stared at the ash mountain that moments ago had been Spark and let out another low keening cry. His dragon, his Spark, was gone.

Charlotte gently pulled him back away from the pile of ash. "Billy, I'm so sorry," she said. Her face was pale, and her eyes were red-rimmed. Billy realized that she was crying and so were Dylan and Ling-Fei.

"She saved us, Billy," Dylan choked out. "She sacrificed herself for us."

"I knew she was good deep down, I knew it," said Ling-Fei, wiping her eyes.

Billy looked up at the crowd of humans and dragons who were still gathered. They were all staring at him as if in a trance, even the copper-colored nox-wing.

A human man Billy didn't recognize stepped forward. "Tell me, boy, why do you cry for the demise of Death's Shadow? She was a terror." He didn't sound angry, just curious.

"I knew her before she became a monster," said Billy. He raised his voice. "She was my dragon, and she was good. The Dragon of Death corrupted her, just as she's corrupted so many humans. It doesn't have to be like this. Dragons and humans can live in peace."

"Why would we want that? Humans are where they belong—at the bottom," hissed a turquoise nox-wing.

"Because it's better for everyone!" Something in Billy snapped. Spark was gone. He felt as if he had nothing left to lose. "Don't you see? The Dragon of Death will control you all! If any dragon ever grows in power, she'll wipe them out. She'll wipe you all out! All she cares about is power. She doesn't care about any of you." His voice broke. "There's a better way."

He braced himself, waiting for a nox-wing to attack him for saying such things about the Dragon of Death, but instead some nox-wings began to nod.

"I would like a heart bond," said one quietly. "That is why I came to the tournament. I know we have been taught to despise humans, but I have never wanted to hurt them."

"I did not think a human would ever cry for a dragon," said another. "Let alone Death's Shadow!"

"That is because this was a true bond," boomed a voice from overhead. Everyone turned to look, and to Billy's shock, he saw that it was Da Huo, JJ on his back. JJ turned his face away from Billy, but not before Billy saw that he, too, had tear streaks down his cheeks.

Da Huo stared down at the pile of ash that had once been Spark and closed his eyes for a long moment. When he opened them, they flashed gold for an instant, before fading back to a fathomless black.

"But even a true bond can be severed by death. The Great One has decreed that this ash will stay here, to remind everyone what happens when anyone—dragon or human—dares to cross her. All hail the Great One! All hail the Dragon of Death!"

The crowd took up the chant. Slowly at first, as if they were still clinging to the hope that there could be another way. But soon the cries echoed all around them. "All hail the Great One! All hail the Dragon of Death! All hail the Great One! All hail the Dragon of Death!"

"The tournament will carry on tomorrow! All are required to return. But, for now, go home. As for you," Da Huo turned to Billy, "I have my orders about what to do with you."

Billy stared back fearlessly at Da Huo. He wasn't afraid anymore. He wasn't afraid of anything. The crowd of dragons and humans slowly dispersed, until it was only Billy, Ling-Fei, Charlotte, Dylan, and Da Huo, with JJ on his back. Thunder still lay unconscious, at the edge of the Dragon Court, but he was breathing and alive. Da Huo eyed the old dragon curiously and then lowered his enormous head so close to Billy's face that Billy could smell his hot, putrid breath.

"That is not one of your dragons. Where are they? The ones who came through to this time?" he whispered.

"They escaped," said Billy flatly.

"They would never leave their precious humans behind. Tell me where they are. The Great One will want all of you. If you do not tell me of your own free will, I have other ways of finding out where they have slunk off to—ways that will be exceptionally unpleasant for you."

"Da Huo," JJ spoke sharply. "Enough. Billy says he doesn't know. Let's return to the Tower and tell the Great One that they got away."

"The boy is clearly lying," said Da Huo.

"I'll tell you where one dragon is," Billy said through gritted teeth. "My dragon is right there. A pile of ash that will be blown away by tomorrow. That's where my dragon is."

Da Huo glanced at the mountain of ash that had been Spark and hung his head. "I am . . . sorry for this loss," he said. "Your bond was true. Spark did not deserve to die like this. No dragon does." He sighed deeply. "The Great One will be expecting you to be delivered to her chambers any moment."

"And we'll tell her that they shot their newfound powers straight into your eyes, and by the time your vision returned, they were gone," said JJ.

"She will smell the lie," said Da Huo. "She is most likely watching us this very moment. You know what she does to those who cross her."

JJ's eyes flitted to the pile of ash. "Then we'll chase them. If that's what she wants."

"JJ . . ." said Da Huo. "You know where my loyalties lie."

"And I know we can both be better," said JJ. Billy stared at the boy who had been their enemy, their friend, their enemy, and now a potential ally.

Da Huo seemed to come to a decision then.

"Run," said Da Huo, rearing up on his hind legs and flapping his wings. "Run and do not look back."

Ling-Fei gazed at the huge dragon. "There's goodness in you. I didn't see it before, but I see it now."

"I think we can debate the goodness of dragons another time," said Charlotte. "Come on. Let's go before he changes his mind."

"Charlotte's right," said Dylan. "We've got to get out of here."

"What about Thunder?" said Ling-Fei.

"The nox-wing will be fine," said Da Huo pointedly. "Now run. Before I change my mind."

With one last glance at the mountain of ash, Billy began to run alongside his friends. He didn't look back, not even when he heard Da Huo's battle roar, or when he felt flames licking at his feet.

Billy knew they couldn't outrun Da Huo, but he didn't catch them, and Billy knew the dragon had purposefully let them go. He wanted to be glad about it, but he didn't feel like he'd be happy about anything ever again

CHAPTER 23
TEARS OF THE MOON

Before Billy could even put his hands on the marble orb, it opened for them, as if it had been waiting. They staggered inside, and as soon as the marble sealed shut, Billy collapsed on the floor. He was still clutching Spark's blue scale.

Midnight flew in, her eyes wide and her horns glowing. "Where's my father?" she said. "Why hasn't he returned with you?"

Tank, Xing, and Buttons emerged from one of the corridors, their expressions grave. "What has happened?" said Tank.

Charlotte ran to him, sobbing. "Tank! Everything is awful!"

"Why is she crying like that?" said Xing, sounding alarmed. "Charlotte is always strong, like Tank. She does not cry for no reason."

"It appears you've all been crying," said Buttons. "Dylan, you must tell us what's going on."

"And somebody has to tell me WHERE MY FATHER IS!" Midnight's horns were dangerously red now. But then, as if she had summoned him, the marble opened once more, and Thunder lumbered in. Midnight flew to him. "Father! You're hurt!"

"I am fine, little Midnight," said Thunder. "A mere scratch. My pride is hurt more than anything." He turned his gaze to the four friends. "I am glad to see you are all back safely. When I woke up alone in the Dragon Court, I thought the worst had happened."

"The worst has happened," said Billy. He held his hand out and opened his fist, revealing the shining blue dragon scale. "This is all that's left of Spark."

Buttons gasped.

"No!" said Xing. "It cannot be!"

"She was so strong," said Tank. "She survived having her wings clipped, she survived the fall. How can she no longer be with us?"

"She couldn't survive the Dragon of Death's wrath," said Billy. "She's dead." He slumped against the marble wall and put his face in his hands. "She's dead." Even as he said the words aloud, he still couldn't believe they were true.

"She saved us again," said Ling-Fei. "The Dragon of Death saw us, and she was going to kill us, but Spark stopped her."

Tank growled at Thunder. "You said you would

protect them! They should have never been seen by the Dragon of Death!"

"How was I to know that they would run out in front of the crowd? I tried to protect them, but it is not my fault your humans are reckless!" Thunder looked away. "I am not as strong as I once was. I have not battled another dragon for many years."

"It's not Thunder's fault," said Billy. "I was seen by a nox-wing who wanted to take me to use as life force. The nox-wing knocked Thunder out."

"If anything, we should be apologizing to him," added Dylan. "We did just leave him in the Dragon Court."

"How dare you!" exclaimed Midnight.

"I was perfectly safe," said Thunder. "Battling for a human is common among nox. Nobody would have suspected anything." Then he frowned. "But I saw Da Huo and his human returning to the Tower when I came to. How did you escape them?"

"They let us go," said Ling-Fei softly. "I'm worried they're going to be punished for it."

"Start at the beginning," said Xing, winding around Ling-Fei protectively. "And tell us everything."

As his friends told their dragons about what had happened, Billy curled up on his side and clenched his eyelids shut, trying to block everything out. All he wanted to do was fall sleep so that he didn't have to think about what had happened.

As the low voices of the talking dragons washed over him, he slipped into a deep slumber. But not a dreamless one. Instead, he dreamed first of Spark. He saw her flying high in the sky, so high she almost reached the moon. Then she turned her head and looked straight at him. Her eyes were gold again and her smile was wide.

"I am free, Billy," she said. "Finally free of the noxious dark magic that controlled me. I will always be with you, even now." Then she flew higher into the night sky until she disappeared among the stars.

"Spark, come back! Don't leave me!" Billy cried, reaching skyward. But Spark didn't reply.

Instead, the round, full moon opened one eye and winked at him. Billy was so startled that he fell backward, and then the dream shifted, as dreams do. The next thing he knew, he was falling through the sky and into a sea covered with pinpricks of starlight. In the middle of it all was the reflection of the round moon. As he fell closer to the sea, the tide began to wash out, and, as it did, Billy saw that there was a graveyard of dragon bones emerging from the waves. He pinwheeled his arms, trying to slow himself down so he wouldn't land in the thicket of broken bones.

And then a honeyed voice spoke close to his ear.

"I cry for them all," the voice said. Enormous salty tears began to rain down from the sky, past a still falling Billy, before they landed on the uncovered dragon bones. More and more tears fell until the tide was high again. Billy looked toward where the voice had come from and saw nothing but the moon, faceless once more. And then he fell into the sea with a splash.

Billy woke with the taste of salt on his lips.

He glanced at his hands and blinked. They were wrinkled, as if he'd been in water for hours. Inside the marble orb, it was impossible to tell what time of day it was because there were no doors or windows. He guessed that it was still night, or perhaps very early morning, because Ling-Fei, Charlotte, and Dylan were all still sleeping. He smiled when he realized that they had positioned themselves protectively around him and someone had put his cloak over him.

He didn't see any of the dragons anywhere, and for that he was grateful. The dragons were light sleepers, and he didn't want them to see what he was about to do—what he had to do.

Billy carefully stood up, stepping over his sleeping friends, and walked across the cavernous room to the other side. He gently placed his hands on the marble and pressed his lips to it in a whisper. "Please let me out. There's something I have to do."

Silently, the marble slid open, just enough for him to slip out. "Thank you," he whispered again. Then he pulled up the hood on his cloak and stepped outside.

The sun was just starting to come up as Billy made his way through the empty streets of Dragon City. The sky itself was a dusty lavender and the gold of the sunrise stretched through it like shining threads in a quilt. With a small smile, Billy saw that the moon was still up, and it was full, just as it had been in his dream.

As Billy pondered, staring up at the sky and lost in his thoughts, a gentle but insistent breeze pushed at his back, urging him onward. The breeze smelled like the sea, like his dream.

When he looked back down at the street, he didn't notice that the moon had started to dip lower in the sky, coming closer, watching to see what he would do.

Spark's ashes were still there, where Billy knew they'd be. Some had blown off in the night, gently dusting the surrounding area in a fine powder, almost like snow. Billy reverently came as close to the mountain of ash as he could, still clutching the single blue scale. He stared down at the last remnants of Spark.

Memories of her, of them together, flashed through his mind in bright and vivid color. He felt warmed throughout, but it was bittersweet, the memories tinged with sorrow now that he'd lost her. Despite this, he wanted to remember her, so he thought back to the first time they flew together. It was like riding the fastest, tallest wave he could imagine, but even better. She'd always reassured him. She'd always protected him.

She'd done it all for him, he realized. Joining the Dragon of Death had been driven by her hunger for dark magic, but her first taste of it, swallowing that star, had been to protect him. But Spark, the strongest creature he knew, hadn't been strong enough to fight off the toxic, noxious draw of dark magic and ill-begotten life force. No matter what, though, she'd always be his dragon in his heart. He would always miss her, always love her.

"Goodbye, Spark," he whispered, tears beginning to slide down his cheeks. "I forgive you." His tears fell faster now, dripping into the pile of ash.

Suddenly, there was a low keening from above, a sound he'd never heard before. Billy quickly turned, expecting to see a dragon about to swoop down and scoop him up, but there was nobody. Nothing except the moon, still visible in the brightening sky. Yet it wasn't just visible, it was closer than Billy had ever seen it before. Then the moon began to cry.

Huge sparkling tears slipped from the moon's round face and down through the sky, mingling with Billy's own tears in Spark's ashes. Billy stared in astonishment at the moon as it continued to weep above him.

Even the moon is mourning Spark, he thought.

There was a rustle at his feet and Billy turned his gaze back to the mountain of ash. The entire thing was quivering, as if there were an earthquake below it. Billy scrambled back, suddenly wary.

A burst of light, golden as the sun, shone through the ash. Then, in one great movement, something emerged from within the mountain of ash and rose up into the sky.

CHAPTER 24
A NEW DAWN

Billy staggered back, holding his hand up to shield his eyes from the dizzying whirlwind of gold and blue brilliance swirling above him. A shimmering blue dragon with great gold wings landed before him and bowed its long neck until its head was at Billy's feet. Billy began to tremble. The dragon raised her head until she was staring directly into Billy's eyes.

"Billy," she said, gold eyes glowing. Billy heard the voice in his ears and in his heart.

"Spark?" his voice came out a broken croak. "But how? You . . . I saw you die . . . and then the moon . . . your wings!" His mind was spinning so fast he couldn't form full sentences.

She smiled at him, and Billy knew, in that moment, that Spark truly had returned.

"We must fly from here. The Dragon of Death must not know what has happened."

"But what about the moon?" Billy said, pointing at the now fading moon. It was far above them now, much too far to have cried tears for Spark.

"The moon has returned to her perch," said Spark. She gazed up at it. "Her tears, and yours, are what brought me back to life." She stretched her gold wings. "And your forgiveness."

As Spark stretched out in her new body, Billy gasped. "Your shadow! It's gone!"

"Good riddance. My shadow had taken on a life of its own—one that was separate from me. It was where all the evil and dark magic I consumed went. My own body felt like nothing but an empty shell. The only flicker of life, of goodness in me, was our bond. It is what kept me from going full nox."

"And now?" said Billy tentatively. He so badly wanted to believe in Spark again, to believe that she was back, truly back. He wanted her to be the Spark he knew and trusted, but she had betrayed him before, and he couldn't let that happen again. What if this was all an elaborate trick plotted by the Dragon of Death?

"Look inside my heart, Billy. You will see no dark magic there, only regret for what I have done and gratitude that I can now fix it."

"Look inside your heart?"

"Yes, through the bond. Close your eyes and focus on seeing what is in my heart."

Billy closed his eyes and focused on Spark's heart. He felt peace, goodness, hope, and joy radiating from it. And a fierce determination that burned molten bright. It felt like Spark again, the Spark he had known, but an even brighter, stronger version.

He opened his eyes. "It really is you."

"I will never let you down again," she said. "We will fix this mess I have made." Then she smiled at him. "Let us fly."

Spark flew up into the sky with Billy on her back. He couldn't believe that Spark had come back, and he was riding his dragon once again.

"Which way?" she called out.

"Toward that marble orb."

"I have long been curious about that orb," said Spark. "It radiates with a strange power, but none comes to the Tower."

"You'll see," said Billy as they landed on top of it.

Billy knew that if the marble orb didn't want to allow Spark in, if it sensed that she was a threat to Thunder, Lightning, and Midnight, it wouldn't open for them. As he slid off Spark and pressed his hands against the marble, he whispered a plea. It was a plea to the marble itself, but also to the universe. A plea that Spark truly was back and that he wasn't letting his loyalty to her lead them all into an even darker future of despair.

The marble opened under his hands, and it kept opening until it was big enough for both of them. Billy turned to Spark. "The others . . ."

"They will be suspicious and rightly so. And they will have questions. But do not worry. All will be well."

They slipped into the marble orb and landed right in the middle of where Tank, Xing, Buttons, Thunder, and Midnight all stood.

They stared stunned at the new dragon in their midst. "Spark? Can it be?" Tank's low voice rumbled with a mix of shock and wariness.

Before any of the other dragons could make a move, Spark opened her new wings wide, sending out a current of power. But it wasn't dark magic, it was something else—something that filled Billy with strength and hope.

Xing gasped. "That is moon magic. It can only be given freely. I have not felt magic like that since I was a young hatchling."

"Billy woke the moon," said Spark. "Her tears—and his—brought me back. And there is no nox in me now. I have left it in the ashes of my former life." She gazed at them. "I hope you can all forgive me, though I know it may take time."

The other dragons stared silently back at her. There was a loud scoff from the edge of the room. Charlotte sat up from where she had been sleeping next to Ling-Fei and Dylan. "*Forgive* you? We're in a terrible future where humans and dragons are *literally* harvested for energy. Have you ever been to the farms? I hope you apologize to everyone there too!"

Dylan cleared his throat. "I think Charlotte speaks for all of us."

Spark hung her head in shame. "I have seen the farms. I have seen the factories. I have seen it all, but I closed my eyes to it."

"Your eyes might have been closed, but your claws were out," said Buttons. "I worry that we no longer know you. You're a stranger to us now. Or maybe even worse than a stranger—an enemy."

"You caused so much pain for so many," added Ling-Fei.

"I know. And I will do everything in my power to fix it. I want to change the future and the past so that it is better one for all," said Spark. "I will show you that my heart is true."

"How can we trust you?" Charlotte crossed her arms and narrowed her eyes at Spark.

Billy couldn't say silent any longer. He had to defend his dragon. "Guys! It's Spark, but . . . a better Spark. She came back without her craving for dark magic. She died protecting us, remember?"

"Technically, it's her fault we were about to be killed in the first place," said Dylan. "Not that I'm holding a grudge or anything."

"Billy, no offense, but you're a terrible judge of dragon character when it comes to Spark. She tricked you before." Charlotte stood and crossed the chamber until she was by Tank's side. "It isn't up to you. We all have to decide if we can welcome Spark back into our clan."

Ling-Fei gave Billy a sad smile. "I want to believe you and I want to trust Spark, too, but she might betray us again." She went to stand by Xing and the long dragon nodded her head in approval.

"Sorry, Billy," said Dylan with a small shrug. "I'm with the girls on this one. You're my best friend and you know I like Spark . . . er . . . or I did, but this is too big of a decision to make based on a feeling you have." He went to stand by Buttons, who patted him on the shoulder.

Billy now stood on one side of the room with Spark, and his friends and their dragons stood on the other side. Midnight and Thunder stood in the middle, eyes darting back and forth.

"I can't believe Death's Shadow is in my house!" Midnight said with a hiccup, ducking behind Thunder and peering out so she could still stare at Spark.

"She's not Death's Shadow! She's Spark!" Billy cried out. "We need her! We're stronger together."

"It is all right, Billy," Spark said softly. "They have every reason to be angry. They need time."

"But we don't have time to waste! The Dragon of Death is still powerful. She's still in charge. We have to stop her before she causes any more destruction," said Billy. He felt overwhelmed with frustration and disappointment that his friends couldn't see what he could. Having Spark on their side was the only way they were going to win!

"I do not know if the Dragon of Death can be defeated in her own destiny," said Tank. "She has amassed so much power. She was already immortal and now she is unstoppable."

"Not quite," said Spark. "She can be defeated, and she knows it. Why do you think she would not allow any heart bonds to form? She knows that a heart-bonded human and dragon are a force to be reckoned with. She can control the dragons and the humans separately, but if they were to be strengthened by the bond, it would be a different story."

"So Xing was right," said Billy. "The tournament was a trap all along."

"Of course I was right," said Xing. "You should never doubt me." She sharpened her gaze toward Spark. "I do not sense any dark magic from you, but you hid it from me before."

When Buttons spoke, his voice was heavy with sadness. "We were in the dungeon, Spark. We had our life force drained by the Dragon of Death."

A gold tear slid down Spark's face. "I will do everything in my power to make it up to you."

"I have a question," said Thunder. "Without the strength of the human and dragon heart bond, how else can we defeat the Dragon of Death? Your human, Billy, is right—time is of the essence."

"We need more dragons," Spark said. "The more dragons we have on our side, the better our chances. Even the Dragon of Death cannot defeat all the dragons in the city. The city may belong to her, but the dragons do not."

"We will never convince all the nox-wings to turn on her," Thunder scoffed.

"We will if they know we can win. This future has not been good to dragons, not even nox-wings. The Dragon of Death's hunger is bottomless. She will drain everyone and everything eventually, and all that will be left is her and Old Gold and an empty world. But we can stop her if we band together and convince the others."

"Wait a second!" Midnight shouted. She was staring at Spark with bright eyes. "Of course! You're the lightning dragon! You can wake my mother! We don't need to hide her power any longer. She can join us in the fight." Midnight was so excited by this prospect that she began to skip in circles around Thunder, her horns alight with joy.

"If you can wake my partner, we will join your cause," said Thunder gravely.

"Is her power the one that is radiating throughout this orb?" said Spark.

Thunder nodded, his long beard and whiskers brushing the ground. "Yes, she generates more power than any dragon I have ever encountered, except perhaps you. We knew if the Dragon of Death sensed her, she would take her away."

"I understand why you have kept her hidden," said Spark.

"She is the one who insisted on it. She put herself to sleep and told us to only wake her when there was another dragon who equaled her in power and who wanted to fight against the Dragon of Death."

"That time has come," said Billy with conviction. "Spark is going to wake up Lightning and then everything is going to change for the better."

•

Midnight and Thunder led Billy, Spark, and the others into the heart of the orb, where Lightning lay sleeping. As they crowded into the space, Spark gasped at the power and energy that Lightning emitted with a single slumbering snore.

"I have never encountered a dragon with so much natural power," she said. "Not even the Dragon of Death. She has to steal all of her power from others."

"Can you wake her up?" said Midnight nervously, fluttering around.

"I can try," said Spark.

"You can do it," Billy said encouragingly.

Spark closed her eyes and took a deep breath. Everyone in the room went very still, except for Lightning who continued to snore. Spark's antlers began to crackle, and her eyes glowed a brighter and brighter gold. She spread her wings and flapped once, twice, three times, as if she were powering herself up. She opened her mouth and shot out a beam of entwined blue and gold light, straight at Lightning's heart

CHAPTER 25
A CHANGED HEART

Lightning's entire body convulsed and then it lifted up into the air as if she were floating. Her eyes snapped open. They were an iridescent, shimmering green one moment, dark blue the next, and then pale yellow, shifting just as her scales did. She grinned a wide smile.

"My baby has grown so much!"

She gracefully landed on the pile of jewels where she had been sleeping. Midnight vaulted herself at Lightning. "Mama!" Lightning nuzzled the young dragon.

A sudden lump formed in Billy's throat, and he desperately missed his own mother, his family, and his home. That was what they were fighting for, he reminded himself, to be reunited with their families. Sensing his thoughts, Spark put a wing around him. "You will see them again, Billy."

Billy managed a shaky nod in return.

Thunder lumbered over to Lightning. "It is good to see your eyes," he said softly. Lightning pressed her snout to his briefly and then lifted her gaze to the other dragons.

But the dragons weren't who she was curious about. "Who are the young humans?"

"It's too much to explain," Dylan said. "But, er, basically, we're all on the same team and we're about to take down the Dragon of Death once and for all, but we need your help."

Lightning's eyes landed on Spark. "And you are the one who woke me?"

"Yes. We are in great need of you and your power."

"Well, what is our plan?" Lightning said expectantly.

"Ah, well, you see, we hadn't got much further than waking you up," said Billy, feeling a little sheepish.

"It sounds like we have much to discuss," said Lightning.

Hours later, when they were still plotting the best way to take down the Dragon of Death, there was a sudden pounding on the roof of the orb. Thunder froze. "The nox-wings. They have come! Your power must be radiating out already. We have to get you to safety!"

Lightning shook her jeweled head. "No, I will not hide any longer." She caught Spark's eye. "I will fight with my new dragon friends by my side."

Spark bowed her head respectfully. "It would be an honor."

"And what about you three? Or shall I say, you six?" Thunder nodded at Xing, Tank, and Buttons and their respective humans. "Will you join us?"

"I will be keeping a close eye on you, Spark," said Xing, but her eyes sparkled. "You are lucky we have known you so long and know your true heart."

The orb shook again with the force of the intruding nox-wings. "We're safe in here, aren't we?" said Midnight, gazing up at the ceiling nervously.

"It depends how many they are. This orb is enchanted, and your mother is strong, but they may be able to break through," said Thunder.

Right then, a crack appeared in the marble above their heads.

"Children, move quickly! Get on our backs and prepare to fly," said Tank.

"Midnight, stay close to me," said Thunder. He looked to Lightning. "I wish our reunion could have been a more pleasant one."

Lightning smiled. "I am glad to be awake. And I am ready to fight."

The marble cracked open with a groan, and a great blast of Lightning's power, which had been stored inside the heart of the orb, shot out into the night air. Suddenly, the sound of dragon roars echoed down into the orb.

"There are so many of them," Billy whispered.

"Do not be afraid," said Spark. "Together, we are strong."

"And with us, you will be stronger," said an unexpected voice from above.

Da Huo thrust his head into the crack in the marble, knocking his head against the edges to open it wider.

"I see we meet again, old foe," said Tank with a roar. "This will be the last time."

"Wait!" shouted a small figure from behind Da Huo's head. "We want to help!" It was JJ.

The dragons inside the marble orb stared at each other.

"It's a trap," said Charlotte. "It has to be!"

"Didn't you hear Da Huo? You'll be even stronger with us!" said JJ. "We've gathered the nox-wings who aren't happy with how things are in Dragon City. When the Dragon of Death clipped Spark's wings, well, it was too much for a lot of them. If she could do that to Death's Shadow, she could do it to any of them."

"She keeps all the power for herself!" shouted a voice from farther back.

"She sent my entire clan to the farms!" cried another.

"How many of you are there?" said Xing.

Da Huo's mouth split into a smile. Billy never thought he would be glad to see the once evil orange dragon grinning. "Come up and see."

Spark looked at Xing, Buttons, and Tank. "A heart can change," she said softly. "For good or for worse. Mine went nox and back. Da Huo was once a friend. He could be again."

"He's the one who brought back the Dragon of Death in the first place," said Buttons gravely. "I fear his loyalty may still be to her."

"Buttons, try to trust me, old friend," said Da Huo. "My human tells me how he flew on you, how you kept him safe before he found me and we bonded. For that, I will always be grateful."

"I trust them," said Billy. "Da Huo and JJ and all the nox-wings they've brought."

"What if it's a trap?" said Dylan.

"Then they will learn very quickly that we cannot be caught," said Lightning. "I have had a long rest, and I would like to stretch my wings."

Billy began to grin.

"Get back!" called Thunder, and the crowd of dragons on the roof of the orb lifted up. With an ear-piercing roar, Thunder split the marble orb in two.

Billy gasped. Above them were hundreds of dragons. He even recognized some of the former nox-wings, like the copper-colored one who had tried to take him for life force, and the turquoise one who had chased them on their first night.

JJ was the lone human. He gave Billy, Ling-Fei, Charlotte, and Dylan a salute. "It looks like I'm not so evil after all," he called down.

"We'll see about that," said Charlotte, but she was smiling.

Tank, Xing, Buttons, Thunder, Lightning, Midnight, and Spark rose up into the fray of dragons. The excitement was palpable, flickering in the air.

"What about Old Gold?" Billy asked JJ. JJ's face fell.

"I don't want to be like him," he said somberly. "He's gone too far. He can't come back from all the bad things he's done." JJ looked down. "Just because I've got his blood running through my veins, that doesn't mean I'm evil like him."

Ling-Fei nudged Xing closer to JJ and Da Huo. "We always have a choice, and today you're choosing to be good."

JJ nodded sadly, but then his mouth lifted in a small smile. "Thanks, Ling-Fei."

"So you're really on our team now?" said Billy.

"As long as you stop saying cheesy things like 'on our team,'" said JJ, rolling his eyes.

"What now?" cried out one of the dragons flying next to them. More and more dragons were soaring out of their lairs now, ready to join the fight.

"Now we take back what the Dragon of Death stole from all of us," said Spark. She lifted up into the air above the crowd. "To the Tower!"

CHAPTER 26
STORMING THE TOWER

Billy and Spark led the charge to the Tower. Spark's gold wings glowed faintly in the moonlight, like a beacon in the dark. Lightning flew next to her and just behind were Tank, Xing, and Buttons, all with their humans on their backs. Close by were Da Huo and JJ, followed by Thunder and Midnight, and beyond that were hundreds of roaring dragons, ready for revenge and retribution.

Below them, the streets of Dragon City filled with humans who had come outside to see what was happening. As they approached the Tower itself, Billy spotted the silhouette of the Dragon of Death curled atop the spire. Old Gold floated on his cloud next to her, waiting for them. Even though they were legion, and she was only one dragon with nobody but Old Gold by her side, Billy felt a shiver of dread begin to unspool inside of him.

Suddenly, the Dragon of Death shot up into the air in an explosion of purple and green light. Her voice boomed all around them, reaching into the ears of every dragon and human in Dragon City.

"What is this? So many of my dragons. Is this some sort of parade? For me? You really shouldn't have."

At the confidence in her voice, Billy's shiver of dread grew into a full cloak of fear.

"Da Huo, I am especially disappointed in you," the Dragon of Death went on. "I should have expected revolt from my wretched Death's Shadow, but you? I thought you knew the meaning of loyalty."

"I no longer serve you!" roared Da Huo. From his back, JJ held up a fist in solidarity. Then he locked eyes with his grandfather, Old Gold.

"YeYe! Join us!"

But Old Gold shook his head. "The Dragon of Death is my true heart bond. I can't turn on her any more than I can turn on you. I won't abandon her, but I won't hurt you, either."

"I will not make you the same promise," hissed the Dragon of Death.

"My grandson will be safe," said Old Gold.

"He has made his choice, and I make the decisions," said the Dragon of Death.

"JJ! Don't fight this battle," cried Old Gold. "Leave!"

"Your grandson is weak, like Da Huo. Only you and I have the true strength of heart to do what has to be done." Then the Dragon of Death raised her voice even louder so all could hear. "Because I am a just ruler, I will allow you all one more chance to remember who you are. Remember who I am. All fall to the ground and bow before me as your one true ruler for now and forever."

"Never!" shouted Spark, and the dragons all around her took up the chant. "Never!"

"Attack!" roared Lightning, and she sent out a blast of energy directly at the Dragon of Death. Spark joined in and dragons from all sides began to swoop and dive toward the Tower and the Dragon of Death, shooting fire and ice and whatever power they possessed.

But the Dragon of Death easily dodged the attacks and then she laughed, the cold sound echoing in the night. "Your feeble attempts are whetting my appetite for a bigger taste of what you all have to offer." Her eyes flashed. "I will show no mercy." With an earth-shattering roar, she flapped her wings and sent out a blast of billowing purple noxious smoke that rushed toward the attacking dragons. Small tentacles shot out of the smoke, each one heading toward the heart of a different dragon.

Billy felt Spark falter beneath him. "That is the strongest nox-cloud I have ever seen. It will drain our life force and weaken us until we are completely unable to fight." She looked to Lightning. "But we have our own cloud of might!"

Lightning nodded, and together with Spark, they flapped their wings back at the Dragon of Death and her noxious cloud, sending out two clouds of their own. Spark's crackled with vibrant blue and gold electricity and Lightning's glimmered in jewel tones. The clouds combined together and hit the noxious cloud head-on.

Billy could feel how exhausting it was for Spark to use such an enormous amount of power. He sent as much of his energy down their bond as he could.

"We'll hold off the cloud as best as we can," he yelled to Ling-Fei, Charlotte, and Dylan. "All of you need to try to get to the Dragon of Death!" But then he saw that his friends had gone limp on their dragon's backs, their eyes wide and vacant. The noxious cloud was slipping past Lightning's and Spark's defense, and it was already taking life force.

"No!" Billy screamed. "This can't be happening!"

Even Tank's mighty wings were beginning to droop, and behind him, Thunder faltered in the sky, barely able to stay upright. Midnight darted around him, and they both flickered in and out of visibility. But her blanketing power could only hide them, not protect them from the onslaught of noxious power that the Dragon of Death continued to attack them with.

As she drained life force, the Dragon of Death began to grow larger and larger, power radiating from her. "You are all fools! And I will drain you all for everything you are worth. Never forget, you are worth more in life force! You are NOTHING!" As her voice rose, Spark and Lightning kept sending out their own blasts of power, but it was futile. Billy felt something break inside of him. They were going to lose.

Nothing could stop the Dragon of Death.

CHAPTER 27
THE END OF EVERYTHING

This was it. The end.

They had fought so hard. They had come so far. And yet the Dragon of Death was still going to win. Billy wished he was closer to his friends so they could be together for their final moments. He tilted his head back to gaze up at the sky one last time. The moon stared back at him.

The moon. The moon!

"Spark!" Billy shouted, suddenly rejuvenated. "The moon! The moon helped you before, can't it do something now?"

"The boy is right!" Lightning cried back, still flapping her wings and sending out her cloud of power to try to slow down the noxious cloud. "The moon is more powerful than she seems, and she helps those who deserve it."

"And now we need her help," said Spark.

"The dark has ruled for too long," said Lightning.

"We'll have to be quick," said Spark. "Without us down here, the noxious cloud will spread even faster. But it is our only chance."

"If we fail, all will be lost," said Lightning.

"If we do not try, we have already failed," said Spark.

"Can you do it on your own?" asked Lightning. "I can try to hold her off by myself."

"I will not be on my own," said Spark. Then she spoke to Billy. "Billy, hold on to me tighter than you ever have, and do not let go. I need your strength and your goodness to help me harness the moon."

"*To what?*" exclaimed Billy. But Spark was already shooting high into the sky, higher than Billy had ever flown. They flew so fast that everything was a blur and soon they had left the Tower and the battling dragons far below them.

Spark flew so swiftly that tears began to stream out of Billy's eyes and his hands grew cold. Up and up they went, the sky got darker as they rose, the stars shone more brightly, but the moon still sat so far out of reach. Then Spark reared up in the night air and sent a shining rope of light, like a lasso, toward the moon itself. The rope landed true. The moon shimmered and Billy could have sworn an eye opened and gazed at them kindly.

"Turn!" Spark cried to the moon. "Please turn!"

Very slowly, the moon began to turn like a dial in the sky.

Billy held his breath. "Is that the dark side of the moon?"

"We never see the dark side of the moon," Spark said, "and we never will. We only see our light reflected back at us. But the moon has her own power—*she* can turn things, like the tides. Now it is time for the Dragon of Death's ill-begotten power to turn back to where it came."

With one more pull, the moon completed a full rotation, and Spark dived back down to earth. They plunged through the clouds and into the fray of dragons. In the center of it all was the Dragon of Death, but she wasn't moving. Her body was still, and she was floating, as if in an invisible sea.

Then she began to violently shake, and before Billy's eyes, the noxious cloud changed from the evil purple he associated with the Dragon of Death to a faint glowing yellow. It was almost the exact shade of the moon overhead. A gust of wind blew streams of the crackling cloud at the amassed dragons and as it washed over Tank, Xing, and Buttons, they inhaled deeply. Billy saw the light come back into their eyes, and at the same time, Ling-Fei, Charlotte, and Dylan sat up and looked around with dazed expressions on their faces.

The power and life force that the Dragon of Death had just stolen from everyone around her started to return to where it had come from, and she began to shrink.

"To ground!" roared Spark. "To ground!" She dashed to the base of the Tower, right as the Dragon of Death dropped down out of the sky. Tank, Xing, and Buttons followed quickly, as did Da Huo, Lightning, Thunder, and more dragons than Billy could count.

But there was only one dragon he had his eyes on. The Dragon of Death writhed on the ground of the Dragon Court as more and more power leeched out of her.

"What's happening?" yelled Charlotte. "Should we attack?"

"Not yet. The power and life force she has stolen is going back to whom it belongs—to where it belongs," said Spark. "To humans. To dragons. To the very earth itself."

"All of it?" said Ling-Fei.

"All of it."

The Dragon of Death screamed as more and more of her ill-begotten power and life force left her. It was going faster now. Billy could hear dragons around him sighing in relief as they gained back their life force. Farther in the distance, he could hear cheering from the factories and the farms as those there regained their strength.

The Dragon of Death looked right at Billy and his friends. "You are fools," she said in a voice so weak Billy almost didn't hear her. "Fools! Do you not know I am the thing holding this destiny together? This future? Without me, it will all crumble and fade to nothing. You will all be nothing!"

Finally, all that remained of the Dragon of Death's former glory was a small dragon with huge fragile wings, glaring at them with furious eyes. Old Gold had landed next to her at some point, and he now looked like a tired old man, not an all-powerful sorcerer. But he stood close to his dragon. "I will be loyal to her until the end," he said. "Goodbye JJ."

"YeYe!"

"I must go where she goes."

"But *where* is she going?" said Dylan. "What do we do now?"

"Remember once upon a time when we first caught her, the plan was to send her into the sky as a dark star and then into a black hole, where no light could find her? That is still what we intend to do," said Xing.

"Now!" cried Spark. Using their combined strength, the dragons surrounding the Dragon of Death blasted her with power, shrinking her further until she wasn't even a dragon at all. Billy gasped as a dark, shining, pulsing star rose up into the sky, a smaller one following in its wake. Now all that was left of them was Old Gold's staff.

The two dark stars rose higher and higher and then shot off into the universe, like shooting stars in reverse, before completely disappearing.

All was silent.

"It is done," said Spark. "She is gone forever."

The Dragon Court erupted in cheers and joyful dragon roars. Billy ran to Ling-Fei, Charlotte, and Dylan. They all jumped up and down and hugged. "We did it!" cried Charlotte. "We finally did it!"

"I'm so happy I never have to see the Dragon of Death again," said Dylan. "So, so happy. All I want is a quiet and peaceful life with some snacks on the side."

"I'm sure that can be arranged," said Billy, slapping his friend on the back. Then he saw JJ standing next to Da Huo, looking drawn and tired.

"JJ! Come here!" he yelled. "We've saved everyone!"

JJ smiled weakly and began to walk toward Billy and the others.

But then the ground beneath them began to shake, making it impossible to walk. Billy and his friends clutched each other.

"What's happening?" Dylan yelled. "I said I wanted a quiet life! I don't want any surprises!"

"Look at the sky!" cried Ling-Fei.

The sky was fading before their very eyes. It was disappearing, crumbling, just as the Tower above and the earth beneath them was.

"She was right!" said Billy with sudden realization. "This future can't exist without her! We're all going to disappear!"

"I will not let that happen," said Spark. She looked at Tank, Xing, and Buttons. "Do you remember how to make a portal? Now that we have destroyed the Dragon of Death, we can make a portal without fear of her following us. But we do not have much time. We have to work fast."

"With the help of the others, we can do it," said Xing. She looked over to the dragons who had gathered to defeat the Dragon of Death. "Will you help us?"

"What will happen to us?" asked one.

"I do not know. But I know this world will crumble without the Dragon of Death to anchor it. This is our only chance of survival," said Tank.

Several of the dragons came closer, focusing their powers along with Xing, Tank, Buttons, Spark, and the Thunder Clan. Suddenly, a swirling multicolored portal opened up on the ground. It was the only stable thing Billy could see.

"But where will it take us?" he said, staring at it anxiously.

"Children, when you jump in, focus as hard as you can on your own time and your love for your family and friends there. Don't think of us, don't think of dragons. Think of home," said Buttons.

"Home," said Billy. Then he looked at Spark. "You're coming too, right?"

"I will try," said Spark. "And no matter what, remember that I am always with you."

The Tower had almost entirely disappeared, leaving a strange and eerie blankness in its wake.

"There is no more time," said Tank. "Is the portal ready?"

"We will not know if it works until we emerge from it," said Spark slowly. "But better to jump into the unknown than to stay in a disappearing future."

"Hold on tight to your dragons," Billy said to Ling-Fei, Charlotte, Dylan, and JJ. "I'll see you all on the other side."

And with one last glance at the crumbling world around him, at the dark future that was now falling apart, Billy, Spark, and the others leaped into the unknown.

CHAPTER 28
THE MOUNTAIN

Billy opened his eyes. Sunlight filtered down through the bamboo trees above him. Below his fingers he felt warm earth. Birds chirped overhead, and in the distance he heard a river rushing. He sat up. Above him, a familiar mountain towered.

Dragon Mountain.

They'd made it back. Or at least, he'd made it back. A lump rose in his throat. Where were his friends? Where was Spark? What if they hadn't made it through the portal? They had to be here with him, they had to be!

There was a squeak next to his ear.

"Goldie! At least you're here." The tiny gold flying pig squeaked again.

A voice came from a nearby tree. "Hello? Can anyone hear me?"

"Dylan! Are you stuck in a tree again?"

Dylan's head emerged from the foliage. "Oh, Billy, thank goodness you're here too! And, no, I'm not *in* the tree, not the way I was. But I did land on one." He grimaced and rubbed his behind. "It was highly uncomfortable." Then his face grew grave. "Where are the others?"

"We're over here!" cried Charlotte. Billy looked up and saw her and Ling-Fei heading toward them, brushing sticks and debris out of their hair.

"I'm here, too, if anyone cares," said JJ, popping up from behind a boulder.

"But where are the dragons?" said Billy. "And I know *where* we are, but not *when* we are."

"I really hope we're back in our own time," said Dylan.

"Children," said a low voice from behind them. "You did it."

Billy whirled around and saw Tank, Xing, Buttons, Da Huo, and Spark emerge from an opening in the mountain. Further behind them, deeper in the mountain, he spotted Midnight, Thunder, and Lightning, who had managed to come through the portal too. Midnight smiled at him and lifted a wing in a salute. Lightning and Thunder nodded toward them, and Billy waved back, feeling pinpricks of tears behind his eyes. He hoped he'd see the Thunder Clan again. He watched as they turned the other way, away from the Human Realm, his realm, and went in the direction of the Dragon Realm.

But the others—the others headed toward Billy.

"You all made it!" cried Billy, running to Spark. Spark enveloped Billy in her wings.

"We did. And it is thanks to you and the strength of all our bonds. You brought us back to safety, Billy."

"What happened to the other dragons?" asked Ling-Fei. "And all those humans?"

"Their stories have not yet been told, but fear not; they will emerge when the time is right. And with any luck, into a better world for all, humans and dragons alike," said Xing.

"But even though you weren't bonded with Midnight and her family, her attachment to all of you carried her here into this time," added Buttons.

"And what time . . . is that exactly?" asked Charlotte, looking around.

"Listen," said Spark.

Billy heard the birdsong, the river flowing, and there, in the distance, something else.

"Billy! Ling-Fei! Charlotte! Dylan! JJ! Where are you? Time for lunch."

"Who is that?" asked JJ.

"I believe it is someone at your camp, looking for you," said Spark.

"And it sounds like you're late for lunch," said Buttons.

"Lunch!" exclaimed Dylan. "Oh, how I've missed lunch!" But then he looked at the dragons and his face fell. "But where will you go?"

"Will we ever see you again?" said Billy.

"Of course you will," said Xing. "You should have known that, but then again, it always takes you a little while to figure things out."

"Will you come into our world?" said Ling-Fei.

"Not yet," said Tank. "Your world is not ready for dragons. But perhaps one day."

"And you are always welcome in our realm. You know where to find us," said Spark.

"What will you do?" asked Charlotte.

"We must see the state of our own realm. Then perhaps we will open more entrances between our worlds so that one day, hopefully one day soon, dragons and humans can live alongside each other in peace," said Xing.

"I like that plan," said Billy. The pig squeaked and flew out of his pocket, landing on Tank's head. Tank sighed heavily. "I suppose this means you're coming with us," he said with a grumble.

"You'll take care of the pig, right?" said Dylan. "We owe that little one a lot."

"I think this pig has proven that she does not need any taking care of and she will outlast all of us," said Xing. "But, yes, we will keep an eye on her."

"Go now," said Spark, leaning down her long neck to nudge Billy forward in the direction of the camp. "We will see you soon."

As Billy, Ling-Fei, Charlotte, Dylan, and JJ began to walk away from the mountain and back to camp, back to the world they had left so long ago, Spark called out after them.

"Thank you, children, for your braveness and for your loyalty—for everything."

Billy turned and waved. He was torn between wanting to run back to Spark and to hurry to the campgrounds.

But then Spark's voice echoed in his mind through their bond. *Do not worry, Billy. We will be here.* And Billy knew that whatever the future held, the Dragon Realm would always be open to them.

The mountain that stood in two realms watched as the human children went in one direction and the dragons in another. If anyone had looked up, they would have seen the faint shape of the moon, even though it was midday, and the sun was bright. And if they had looked closely, they would have seen that the moon was smiling.

Acknowledgments

We are so grateful that so many readers have wanted to continue the Dragon Realm adventure with us. Thank you for reading this book. We hope you loved reading *Dragon City* as much as we loved writing it.

As always, we couldn't send our dragons out into the world without the help and support of our Dragon Dream Team.

To our agent, Claire Wilson, for always believing in our dragons and in us. We'd ride with you into any battle, even against the Dragon of Death, and we know that, with you on our side, we'd always triumph. Thank you as well to the rest of the team at RCW, especially Safae El-Ouahabi and Sam Coates.

Huge thank you to our team at Simon & Schuster UK, especially our dragon Queen Rachel Denwood and the amazing Amina Youseff in editorial, and Laura Hough, Sarah Macmillan, and Eve Wersocki Morris in sales, marketing, and PR.

Thank you to the team in the US at Sterling Children's Books for all their work on the series, especially Ardi Alspach and Blanca Oliviery.

We continue to have the most beautiful books out there, and that is thanks to Jesse Green for design, Tom Sanderson for the lettering, and the absurdly talented Petur Antonsson for the cover illustrations. Thank you for creating such stunning covers for our books! They make readers want to pick them up and meet our dragons.

We are so grateful to booksellers for helping readers find our dragons! We'd like to especially thank Queens Park Books, Tales on Moon Lane, Waterstones, and Barnes and Noble for their support.

We also want to thank all the teachers and librarians who have championed the series and introduced it to their students. And thank you to the bloggers for shouting so loudly about the books!

We love our writing community and feel so lucky to have so many wonderful friends to help us celebrate the good times and cheer us on when we need it. Thank you to Cat, Kiran, Tom, Anna, Kate, Abi, Katherine, Roshani, Krystal, Alwyn, and Samantha.

And, finally, to our wonderful family across the globe. A dragon-size thank-you to the Tsang, Webber, Hopper, and Liu family members for their support and enthusiasm. This book is dedicated to our mothers, so special thank-yous to Virginia Webber and Louisa Tsang. And thank you to Jack, Jane, Cat, Stephanie, Ben, and Cooper for being the best.

And thank you to our daughters, Evie and Mira, who inspire us the most of all.

Dragons Take Flight in this Epic Adventure!

Don't miss the first two books in the
DRAGON REALM
series

Join Billy, Charlotte, Ling-Fei, and Dylan as they and their heart-bonded dragons explore the Dragon Realm, travel through time, and take on the Dragon of Death.

Available everywhere books are sold.